Achilles

IN THE

Quantum
Universe

RICHARD MORRIS

Achilles

IN THE

Quantum

Universe

· · · · · · ·

The Definitive
History of Infinity

HENRY HOLT AND COMPANY · NEW YORK

Henry Holt and Company, Inc.
Publishers since 1866
115 West 18th Street
New York, New York 10011

Henry Holt® is a registered
trademark of Henry Holt and Company, Inc.

Published in Canada by Fitzhenry & Whiteside Ltd.,
195 Allstate Parkway, Markham, Ontario L3R 4T8.

Library of Congress Cataloging-in-Publication Data
Morris, Richard
Achilles in the quantum universe: the definitive
history of infinity / Richard Morris.
p.   cm.
Includes index.
1. Astronomy—Philosophy—History.   2. Infinite—History.
3. Physics—Philosophy—History.   I. Title.
QB14.5.M67   1997   96-49772
530'.01—dc21   CIP

ISBN 0-8050-4779-4

Henry Holt books are available for special promotions and
premiums. For details contact: Director, Special Markets.

First Edition—1997

Designed by Victoria Hartman

Printed in the United States of America
All first editions are printed on acid-free paper. ∞

1   3   5   7   9   10   8   6   4   2

CONTENTS

Preface ix

1 · The Paradoxical Nature of Infinity 1

2 · Infinite Time 17

3 · Infinite Worlds 32

4 · The Infinitely Small 54

5 · Atomic Catastrophe 71

6 · Electrons Have Infinite Mass 95

7 · There Was a Young Lady
Named Bright 120

8 · Singularities 140

9 · Is the Universe Finite, Infinite,
or Imaginary? 158

10 · Infinite Worlds 184

11 · ∞ 203

Index 209

PREFACE

ALMOST AS SOON as I began making an outline for this book, I realized that a great deal of it would involve topics in contemporary physics and cosmology. As scientists working in these fields have sought to understand the origin of the universe and the nature of physical reality, they have encountered the infinite again and again. At the very beginning of the early twentieth-century revolution in physics, scientists found themselves confronted by the infinite. This created baffling problems that had to be solved before further progress could be made. Similar problems have arisen again and again. Scientists have encountered the infinite in quantum mechanics, in Einstein's theories of relativity, and in theories about black holes. For example, if Einstein's general theory of relativity is correct, the matter at the center of a black hole is compressed into a mathematical point, and is infinitely dense.

Cosmologists have found that the ancient philosophers who speculated about the existence of an infinite number of worlds were much too modest. Findings in the new field of quantum cosmology seem to imply that there might exist an infinite number of universes, many of them much like our own. They may be inhabited by beings much like us. Some universes may differ from our own only in small details. In the last few years, the "alternate universes" of science fiction have become part of sober scientific speculation.

But this is not solely a book about modern physics. It is one that tells the story of the attempts that have been made to grasp the idea of "infinity" from ancient times to the present. As I relate this story, it should become clear that the infinite is just as baffling a thing today as it was in the days of Aristotle. The only real difference is that, as scientists have probed deeply into the nature of our universe, they have caught glimpses of the infinite in ways that the ancient Greeks could never have imagined.

I say little in this book about the mathematics of infinity. Many books have been written on this subject, and the concept of infinity as an abstract mathematical entity is something that is well understood. But such theories have no applications in the physical sciences. When the infinite is confronted in the real world, it becomes something that is mysterious and elusive. In such cases, mathematical theories are of little help.

Although certain early Greek philosophers spoke of an infinity of worlds, the first person to examine the concept of the infinite in detail was the Greek philosopher Zeno. In a series of famous paradoxes, he claimed to be able to show that motion was impossible because an infinite series of acts could never be completed. For example, before one could travel a given distance, it was first necessary to travel half that distance, and then half of the distance that remained, and then half of that, and so on. Since the series was endless, it was impossible to reach one's goal.

At first glance, Zeno's paradoxes seem trivial. However, some modern philosophers believe they raise issues that still have not been settled. Though the paradoxes sound simple, they are immeasurably profound. But a preface is not the place to discuss them in detail; for now, I only want to point out that infinity made its first appearance as a puzzling problem more than two thousand years ago.

The infinite is still as much of a puzzle as ever when Galileo equated it with the "incomprehensible." He gave examples of the paradoxical properties of infinite numbers and admitted he did not understand them. But, strangely, Galileo thought the universe was infinite in extent. Galileo, the originator of the scientific revolution that has continued to our time, believed the universe was something that could not be understood.

Galileo was not the only scientist to struggle with the infinite. Numerous scientists, from Isaac Newton to the British physicist Stephen Hawking, have found themselves dealing with the infinite in one form or another. Newton discovered he had to deal with the infinitely small in order to solve problems that arose in the context of his theory of gravitation. But he could never explain just what the infinitely small was. In fact, two centuries would pass before the problem was eventually solved.

In our own day, Stephen Hawking has evolved a concept he calls "imaginary time" in order to avoid the infinities that would otherwise be present at the beginning and end of time. The idea that space was infinitely compressed at the beginning of the big bang can be avoided, Hawking says, and postulates that under certain conditions time can take on the character of a spatial dimension. If it can, he goes on, then the universe had no beginning. Nor were there three dimensions of space and one of time, as there are today. At the beginning, there was no time, only four spatial dimensions.

The infinite makes its appearance in philosophy and literature, too. Even today, we encounter the idea, proposed by the ancient Stoic philosophers, that the universe is destined to pass though an infinite number of cycles, that the same events will be repeated an infinite number of times. Certain authors, such as Jorge Luis Borges, have found themselves fascinated by the concept. In one of his essays, Borges suggested that encounters with infinity convince one of "the hallucinatory nature of the world." The infinite, he says, brings us into contact with the kind of "unreason" that convinces us that the world as we perceive it cannot possibly be real.

It is easy to sympathize with Borges's views. We often find modern scientists confronting problems very similar to those encountered by the ancient philosophers who attempted to struggle with the infinite. It appears that though the lines of battle have shifted, "infinity" is just as mysterious a thing as it ever was. One may disagree with Borges's contention that the world has a "hallucinatory nature"; however, anyone who contemplates the infinite inevitably encounters ideas that have precisely this quality.

For the most part, I have avoided extended discussions of specific philosophical doctrines and of literature here. This book is not a study of philosophical thought. Nor is it a work of literary criticism. On the contrary, it is an account of an often dramatic, millennia-long struggle to come to terms with the infinite. The early chapters do make reference to quite a few philosophers—after all, in ancient times, it was they who struggled to understand the nature of the universe. Today this is more often a scientific endeavor, so I have placed a strong emphasis on science in the latter part of the book.

I have described the events that make up the story of infinity in chronological order. Chapter 1 is an exception; I added it to introduce the reader to some of the paradoxes associated with the concept of infinity. I have also included a brief discussion of mathematicians' conceptions of infinity in that chapter for the sake of completeness.

Achilles

IN THE

Quantum
Universe

1

THE PARADOXICAL NATURE
OF INFINITY

A BASEBALL PLAYER DIES and goes to heaven. He likes nothing better than playing baseball, so God decrees that he shall be allowed to play every day for all eternity. In other words, he will be able to play an infinite number of games. Furthermore, it is ordained that every tenth hit shall be a home run. "George Herman," says God (who likes to address people by their first and middle names), "you will have an infinite number of hits and an infinite number of homers."

God doesn't mean that this player can expect to hit a home run after every set of nine singles, doubles, and triples. That would take all the excitement away. George Herman will sometimes get twenty or even thirty hits before he slams one out of the park. On other occasions, he will homer on consecutive times at bat. In the long run, as he accumulates hits, the ratio will get closer and closer to an exact one in ten.

Now, an infinite number of games is quite a lot. It is obvious that they are going to require an infinite supply of baseballs. Since God doesn't want to be bothered with having to constantly perform miracles to create them, He gives each team a somewhat nondescript wooden bin containing an infinite number of balls. Obviously, the supply can never be exhausted. However many times you subtract one from infinity, an infinite number remains.

George Herman is a very good hitter. But one of the pitchers on the

1

team does not hit too well at all. In fact, it has been ordained that although he will get a great number of batters to strike out, he himself will finish the infinitely long season with a batting average of zero. One day, to everyone's surprise, the pitcher gets two hits. At first, no one can understand how this is possible. After all, it is inconceivable that God would lie to His players. However, one of the other players, a man named Yogi who is something of a philosopher, soon hits upon a solution. If the pitcher only gets a finite number of hits in an infinite series of games, he will indeed wind up with a .000 batting average. If you divide any number by infinity, Yogi says, the result is always the same. Two divided by infinity is zero. If you divide two thousand, or two million, or even two trillion by infinity, the result is still zero. "It ain't over till it's over," he adds. Not all of the players are convinced by this analysis. "How can you divide a number by infinity?" one mutters.

Naturally, George Herman's team plays half its games on the road. And of course when the players travel, they must have a hotel to stay in. They never bother to make reservations in advance—in heaven, all the hotels have an infinite number of rooms.

But one day the team arrives in another celestial town and discovers that the hotel is full. There are already an infinite number of guests. At first it appears it will be necessary to seek other accommodations, but then the archangel who manages the hotel quickly assures the team that this will be unnecessary. Yes, the hotel is full, yet there will be no problem assigning each player a room. The archangel asks how many rooms he must provide and is told that, counting players, coaches, and so on, a total of forty is required. The manager then does a little juggling. He moves the guest occupying room number 1 into number 41. He shifts the guest in number 2 into number 42, and so on. When he is finished, forty rooms are available. Furthermore, no one has been evicted. Everyone previously occupying a room has simply been moved into one with a higher number. The woman who was originally in room 41 is now in room 81, for example.

"That was easy," the archangel said to himself after each of the baseball players has been given a key to his room. "If I had to, I could have accommodated an infinite number of new guests." And indeed he could. He could have moved the guest in room 1 to room 2, while moving the guest in room 2 to room 4, the guest in room 3 to room 6, and so on. This would have emptied all the rooms with odd numbers—and of course there were an infinite number of them.

It appears that infinite numbers—if they can indeed be called numbers—are paradoxical entities. You might legitimately wonder if it is meaningful to imagine a baseball season of infinite length, or to talk of a hotel with an infinite number of rooms, as such things just don't exist in the real world. But before I attempt to address this concern, let's look at yet another paradoxical property of infinite numbers.

It can be easily shown that if George Herman gets one home run for every ten hits, then the number of home runs and the number of hits are equal. To show this, you need only to pair hits and home runs in a one-to-one manner. The first home run is matched with the first hit, the second home run is paired with the second hit, and the nine hundred and ninety-ninth home run corresponds to the nine hundred and ninety-ninth hit. For every hit, there is always a corresponding home run. Since the number of home runs is never exhausted, there are no hits "left over."

Some matters are easier to understand if expressed in a visual form. I will therefore give another example of this kind of pairing of sets of numbers, and show that the number of positive integers (the whole numbers 1, 2, 3, and so on) is equal to the number of integers that can be divided by two. I need only write down the positive numbers in one row, and the even numbers in the row below it, as follows:

$$1 \quad 2 \quad 3 \quad 4 \quad 5 \quad \ldots$$
$$\updownarrow \quad \updownarrow \quad \updownarrow \quad \updownarrow \quad \updownarrow$$
$$2 \quad 4 \quad 6 \quad 8 \quad 10 \quad \ldots$$

Here the double-edged arrows indicate that each number is paired with the one below it, and the three dots at the end of each row indicate that both series go on forever.

At first, this may look like a somewhat suspicious argument. But in fact the reasoning is perfectly valid. Pairing one collection of objects with another is the most fundamental form of counting. It is a procedure that can be employed even if one knows no arithmetic.

This can be illustrated by the following example: Imagine that a young girl wants to know if the number of cups and saucers that her mother keeps in the cupboard are equal. Although the little girl is precocious, she has not yet learned to count. However, this is no hindrance. She simply places each cup in a saucer. If there are no cups and no saucers left over when she finishes doing this, the numbers of each are the same.

Even though intuition would tell us that there are twice as many positive integers as there are even integers, we must conclude that the two infinities are equal. Indeed, it is possible to find even more extreme examples. In 1638, for instance, the great Italian scientist Galileo noticed that the number of positive integers was equal to the number of squares.

A square is a number formed by multiplying any number by itself. The first square is 1, the result obtained when one carries out the multiplication 1×1. The next squares are 4 (2×2), 9 (3×3), 16 (4×4), 25 (5×5), and 36 (6×6). They can be paired in the same manner that numbers were paired above:

$$
\begin{array}{cccccc}
1 & 2 & 3 & 4 & 5 & 6 \quad \ldots \\
\updownarrow & \updownarrow & \updownarrow & \updownarrow & \updownarrow & \updownarrow \\
1 & 4 & 9 & 16 & 25 & 36 \quad \ldots
\end{array}
$$

When Galileo obtained this result, he concluded that there was something very bizarre about infinite numbers, and that they were best avoided. Infinity, he said, was "inherently incomprehensible." He was not the first or only person to come to this conclusion. The fact that "unequal" infinite collections could be paired with one another in this way had been noticed in ancient times, and for a couple of centuries or so after Galileo's death, mathematicians generally denied that it was meaningful to speak of infinite numbers. They did make use of unending "infinite series" of numbers. A simple example of such a series would be the set 1, 2, 3, . . . that we have already encountered. Another would be the series of fractions ½, ¼, ⅛ . . . . But here the basic idea was that such a series, like the Energizer bunny, kept going, and going, and going. Even in a steadily increasing series such as 1, 2, 3, . . . you never actually got to infinity, only to progressively larger numbers.

TRANSFINITE NUMBERS

The concept of infinite numbers continued to baffle mathematicians until the latter part of the nineteenth century. Then, in a series of papers published between 1874 and 1884, the German mathematician George Cantor showed that infinity could indeed be treated in a mathematically rigorous way. He began by *defining* an infinite number to be one that could be put into a one-to-one correspondence with some part of itself. I have already given two examples of this by showing that the positive integers can be matched both with the even integers and with the set of square numbers. This gives us the not very surprising result that, according to Cantor's definition, the numbers 1, 2, 3, . . . constitute an infinite collection.

But Cantor did much more than define infinity. He achieved a number of surprising results. For example, he was able to prove that the set of positive integers had the same number of members as the set of all proper and improper fractions.* To do this, it was only necessary to set up the following one-to-one correspondence:

$$
\begin{array}{cccccccccc}
1 & 2 & 3 & 4 & 5 & 6 & 7 & 8 & 9 & 10 \quad \cdots \\
\updownarrow & \updownarrow & \updownarrow & \updownarrow & \updownarrow & \updownarrow & \updownarrow & \updownarrow & \updownarrow & \updownarrow \\
\tfrac{1}{1} & \tfrac{2}{1} & \tfrac{1}{2} & \tfrac{1}{3} & \tfrac{2}{2} & \tfrac{3}{1} & \tfrac{4}{1} & \tfrac{3}{2} & \tfrac{2}{3} & \tfrac{1}{4} \quad \cdots
\end{array}
$$

Note that the lower series is ordered in such a manner that no fraction will be left out. Cantor begins by including all fractions in which the numerator and denominator add up to 2. There is exactly one of these, $\tfrac{1}{1}$. He then lists the fractions in which this sum is 3. This time there are two: $\tfrac{2}{1}$ and $\tfrac{1}{2}$. Next, there are three fractions for which the sum is four, four for which the sum is five, and so on.

Some of the results that Cantor obtained were surprising indeed. For example, in 1874 he set out to prove that the number of points on a line was less than the number of points on a plane or in a space of any num-

---

* In improper fractions the numerator is greater than the denominator, such as $\tfrac{3}{2}$ and $\tfrac{7}{4}$.

ber of dimensions.* Instead he discovered a proof of the opposite. No matter how many dimensions there were, the number of points was always the same. "I see it, but I do not believe it," he said in a letter he wrote to the German mathematician Richard Dedekind in 1877.

But don't imagine that Cantor's work implied that all infinite numbers were equal. This was definitely not the case. For example, he was able to show that the positive integers 1, 2, 3, . . . could not be put into one-to-one correspondence with the points on a line. This meant that the latter infinity had a greater magnitude. Even though both were infinite, the number of points on a line was larger than the number of positive integers.

Eventually, Cantor was able to show that there were many different infinite numbers—an infinite number of them. He assigned the symbol $\aleph_0$ to the smallest infinity, the one represented by the positive integers.** Here $\aleph$ is the first letter of the Hebrew alphabet, aleph, and $\aleph_0$ is said as "aleph-null." The next larger infinite number is $\aleph_1$, aleph-one, and is followed by an unending series of infinite numbers, all represented by the same Hebrew letter. Cantor called the alephs transfinite numbers, and they are still known by that name today.

As might be expected, Cantor's discoveries did not gain immediate acceptance among mathematicians. Many of them wanted to avoid the use of the concept of infinity altogether, and here was Cantor speaking of an infinite number of infinities. One of Cantor's former professors, the German mathematician Leopold Kronecker, was especially critical of his work. He attacked Cantor's ideas as being "mathematically insane," and later prevented his former student from obtaining a post at the University of Berlin. Another, even more eminent mathematician, the Frenchman Henri Poincaré, described Cantor's mathematical theory of infinity as something that later generations would regard "as a disease from which one has recovered."

---

* Mathematicians frequently speak of "spaces" with more than three dimensions. Indeed, they sometimes make use of spaces of an infinite number of dimensions. But it should not be imagined that there is any relationship between these abstract mathematical spaces and the three-dimensional space of the everyday physical world. The former are purely mathematical constructs.

** $\aleph_0$ also represents the set of all squares, or the set of all fractions. As we have seen, these are equal to the set of all positive integers.

Such attacks had an unfortunate emotional effect on Cantor. Somewhat paranoid to begin with, Cantor began imagining conspiracies. He refused to have anything to do with the only mathematical journal that had been receptive to his work, believing its editor was involved in a plot against him. He experienced a nervous breakdown in the spring of 1884. After recovering, he withdrew from mathematical work and began to publish essays in philosophical journals. During the latter part of his life, Cantor experienced severe depressions and several mental breakdowns. He was eventually relieved of his teaching duties at the University of Halle and ended his life in a mental hospital in 1918.

By this time, a younger generation of mathematicians and philosophers was beginning to understand the importance of the work Cantor had done. In 1926, the great German mathematician David Hilbert summed up the newfound regard for Cantor by saying, "No one shall expel us from the paradise that Cantor has created for us." But of course, by then Cantor had been dead for eight years.

ACHILLES AND THE TORTOISE

It is not my intention to discuss Cantor's theory of transfinite numbers in great detail. The theory is an example of what is called "pure" mathematics—that is, mathematics for its own sake—and has no applications in the natural sciences. Physics, for example, does not make use of transfinite numbers, nor does any other scientific field. My purpose in introducing the topic was simply to show that, however paradoxical the concept of "infinity" may seem, it is nevertheless one that can be put on a firm logical foundation. You cannot simply refuse to admit infinite numbers into mathematical discourse, as Leopold Kronecker wanted to do.

This makes the idea of infinity more, not less, baffling when it is met in a nonmathematical context. When infinite quantities are encountered, we cannot wish them away by saying that infinity is an illogical or self-contradictory concept. We can't dismiss infinite numbers as "inherently incomprehensible," as Galileo did. If we encounter some situation in which infinite quantities appear, it is necessary to examine matters carefully and to try to find a way to work with these numbers.

It is true that infinite quantities are not encountered in the everyday world. Nothing travels at an infinite velocity. There are not an infinite number of stars in the sky, or an infinite number of grains of sand on the beach. However, one encounters the concept of the infinite again and again in philosophy, in modern science, and occasionally in literature. In everyday speech, the word *infinite* is still often used as a synonym for "that which is beyond human comprehension." When the infinite is encountered in a scientific or philosophical context, however, it cannot be evaded that easily. Science and philosophy, after all, are attempts to understand the world.

One of the earliest and most famous uses of the idea of infinity is the paradox of Achilles and the Tortoise, which was conceived by the Greek philosopher Zeno of Elea sometime around the middle of the fifth century B.C. It can be stated as follows: Suppose that the swift Greek warrior Achilles is to run a race with a tortoise. Because the tortoise is much the slower of the two, he is allowed to begin at a point some distance ahead. But then, says Zeno, Achilles can never overtake his opponent. For to do so, he must first reach the point from which the tortoise began. By that time, the tortoise will have run to some point farther down the racecourse. And by the time Achilles reaches *that* point, the tortoise will have advanced farther yet. It is obvious, Zeno maintains, that the series is never-ending. There will always be some distance, however small, between the two contestants.

Of course, we all know that Achilles would catch the tortoise fairly quickly, but pointing this out does not refute Zeno's argument. Zeno is saying that Achilles must complete an infinite series of acts, and this cannot be done in a finite period of time. If we choose not to believe this, we must demonstrate where the fallacy lies.

THE ONE

Before I continue the discussion of Zeno's paradox, it might be a good idea to say something about the context in which it was proposed. When Zeno formulated "Achilles and the Tortoise" and his other paradoxes, he was not simply trying to invent amusing puzzles. On the contrary, his purpose was a serious one. Zeno was a disciple of the philosopher Parmenides, who maintained that reality was an unchanging unity, which he called the One. Motion, change, and multiplicity, said Parmenides, were illusions. Those who believed that they were real were being deceived by their senses. Parmenides' ideas bear a certain resemblance to those encountered in Eastern philosophies. The Hindus, for example, also maintain that the world of the senses is an illusion. There is a significant difference, however. Parmenides believed that rational thought could uncover the true nature of reality; he was not following any "spiritual path."

Parmenides is generally considered to have been the first rationalist; that is, the first philosopher to follow wherever reason seemed to lead him, even if his conclusions appeared to be contradicted by common sense. His ideas seem to have impressed Plato, who wrote a dialogue—called *Parmenides*—which depicts a young Socrates as being much in awe of the older philosopher. As one might expect, Parmenides' ideas were also subjected to a great deal of ridicule. It is thought that Zeno invented his paradoxes in order to defend his teacher.

Unfortunately, Zeno's own writings are lost, and the only account we have of his paradoxes is that of Aristotle, who stated them in order to refute them. Aristotle does this in a hasty and careless manner; it appears he didn't take the paradoxes too seriously, so we cannot be sure we have the paradoxes in their original form. For that matter, it isn't certain precisely what point Zeno was trying to make. Aristotle says that Zeno proposed the paradox of "Achilles and the Tortoise" and another paradox called "The Dichotomy" in order to show that motion was impossible. But it's not certain this is correct. Some scholars think Zeno was arguing against the idea that space and time were infinitely divisible, that his point in describing an absurd situation in which Achilles must run through a series of distances that grow progressively shorter was to show that space couldn't be divided up in this way. Indeed there is a great deal of logic in this point of view. If space could be infinitely subdivided, this would contradict Parmenides' ideas. You could hardly say that something with an infinite number of parts is an unchanging unity.

Aristotle credits Zeno with being the inventor of dialectic, a technique frequently used in the Platonic dialogues. In these writings we often see Socrates asking some other person to state an opinion. Socrates then demonstrates that this idea leads to a contradiction or to an absurd conclusion.* Apparently Zeno was using the technique of dialectic in "Achilles and the Tortoise." Everyone knew that Achilles would soon overtake the slow-footed animal. Consequently there had to be something wrong with the initial assumptions.

Knowing something about the reasons that Zeno created his paradox really doesn't help us to understand it. If we want to accomplish that, we must look at the paradox itself. To make matters a little simpler, I am going to assume that Achilles runs exactly twice as fast as the tortoise. This may sound a little unreasonable, but then maybe Achilles has had a hard day slaying Trojans, and if the tortoise wasn't the fastest turtle in the world, he probably wouldn't have challenged Achilles to a race in the first place. Note that making this assumption doesn't alter the nature of the paradox in the least. The principle is exactly the same whether Achilles runs twice as fast, or ten times as fast, or fifty times as fast as his opponent.

I will assume further that the tortoise has been given a headstart of ten meters, and that it takes exactly one second for Achilles to complete the first stage of the race; that is, to reach the tortoise's starting point.** It is easy to see that the tortoise's lead will be cut to five meters at this point. If Achilles can run ten meters in a second, the tortoise will run at half that speed. Since the tortoise's lead has been cut in half, it will obviously take Achilles only one-half second to complete the second stage. Completing the third will require a quarter of a second, while the fourth will take an eighth of a second, and so on. If I put this in tabular form, we get something like this:

Lap	1	2	3	4	5	6	...
Time required (seconds)	1	$\frac{1}{2}$	$\frac{1}{4}$	$\frac{1}{8}$	$\frac{1}{16}$	$\frac{1}{32}$	...

---

* Dialectic (showing that an idea leads to a contradiction) is not the same thing as Socratic method (a technique of questioning). Socrates often combined the two methods.

** I realize that no one actually runs this fast. My intention is to choose numbers that make the arithmetic easier.

If we then add up the total time that has elapsed at any stage of the race, we find that the sum is 1½ seconds after two laps, 1¾ seconds after three, 1⅞ seconds after four, and so on.

Lap	1	2	3	4	5	6	. . .
Total elapsed time (seconds)	1	1½	1¾	1⅞	1¹⁵⁄₁₆	1³¹⁄₃₂	. . .

It looks as though the total time gets closer and closer to two seconds. Indeed, in the real world Achilles would catch the tortoise in exactly that amount of time under the conditions that I have described. And if Achilles had been running ten times as fast as the tortoise, the result would be similar. The only difference would be that the time required for the race would be less. In fact, it can be shown that the time would be 1⅑ seconds.*

At first, it appears that Zeno's paradox can be disposed of rather easily. All that is required is a little arithmetic. A moment of reflection will show that this is not the case. Zeno did not say that Achilles could not catch the tortoise in a finite time. He knew very well that this was exactly what would happen. What Zeno was really saying was it was impossible for Achilles to perform an infinite number of acts.

THE DICHOTOMY

The paradox called "The Dichotomy" (the names by which the paradoxes are known were coined by later commentators, not by Zeno) is similar in nature to "Achilles and the Tortoise." According to Zeno, it is not possible to complete any journey. In order to do so, you must first travel half the distance to your goal, and then half of the remaining distance, and again half of what remains, and so on. However close you get to the place you want to go, there is always some distance left.

Furthermore, it is not even possible to get started, Zeno says. After all,

---

* Because $1 + 0.1 + 0.01 + 0.001 + . . .$ is a number that gets closer and closer to the decimal $1.1111111 . . .$ (here the dots indicate that the 1's repeat forever), which is exactly 1⅑.

before the second half of the distance can be traveled, one must cover the first half. But before *that* distance can be traveled, the first quarter must be completed. And before that can be done, one must traverse the first eighth, and so on and so on.

The two forms of the paradox are really mirror images of each other. In the first, Zeno splits a distance into increasingly small parts. If we represent this with a series of fractions, we get the infinite set ½, ¼, ⅛, ¹⁄₁₆ . . . . In the second part of the paradox, this is reversed, and the increasingly small fractions come at the beginning: . . . ¹⁄₁₆, ⅛, ¼, ½.

If we add up the fractions, we get something like this: ½ + ¼ + ⅛ + ¹⁄₁₆ + ... As more terms are added, the sum gets increasingly close to 1, just as it got closer and closer to two seconds in the previous paradox. Zeno is saying that to traverse any distance, it is necessary to perform an infinite number of acts.

It is said that when Zeno related this paradox to Diogenes the Cynic, Diogenes "refuted" it by getting up and walking away. But it is safe to assume the story is apocryphal; Zeno died seventeen or eighteen years before Diogenes was born. In any case, it would not really have been a refutation. Zeno knew very well that people were generally able to move about. He was asking how it was possible.

Zeno's paradoxes are not easy to dispose of. Though at first they may seem to be whimsical little puzzles, they become truly mysterious when they are examined in detail. In an essay written during the 1920s, the British philosopher Bertrand Russell characterized them as "immeasurably subtle and profound," and other twentieth-century philosophers have argued about them at great length. These philosophers can be divided into two camps: those who think there is no real problem, and those who believe that Zeno's paradoxes have not yet been solved.* Two and a half thousand years after he first set them forth, Zeno's paradoxes still tantalize some of our greatest intellects.

---

* For the benefit of the reader who wants to delve deeper into these arguments, a number of philosophical papers on the subject are collected in a book edited by Wesley C. Salmon, *Zeno's Paradoxes* (New York: Bobbs-Merrill, 1970).

CREVICES OF UNREASON

The Argentinean author Jorge Luis Borges developed his own fascination with Zeno's paradoxes. In his mid-twentieth-century essay "Avatars of the Tortoise," Borges speaks of the "hallucinatory nature of the world": "We have dreamt it as firm, mysterious, visible, ubiquitous in space and durable in time, but in its architecture we have allowed tenuous and eternal crevices of unreason which tell us it is false." One of the "crevices of unreason" of which Borges speaks is the one that contains the paradoxes of Zeno. Borges was fascinated by that "numerous Hydra"—the infinite—and references to it appear in many of his stories. Borges once remarked that he had long wanted to write a book on the subject, but was deterred by the fact that too long a period of "metaphysical, theological and mathematical apprenticeship" would be required.

A somewhat different approach was taken by the British author G. J. Whitrow in his 1961 book *The Natural Philosophy of Time*. By rephrasing Zeno's fundamental idea, Whitrow succeeds in making it seem even more mysterious than it was in its original form. Whitrow asks us to consider the case of a bouncing ball that reaches three-quarters of its former height on each bounce. Since the ball doesn't rise as high on each succeeding bounce, the time required for each cycle will steadily decrease in exactly the same manner that distances and times decrease in Zeno's paradoxes. The only significant difference is that Whitrow uses a factor of three-quarters where Zeno used one-half. Whitrow shows that, using certain reasonable assumptions about the ball's initial velocity, it is possible to calculate that it will bounce an infinite number of times in the space of four seconds.

It wouldn't make any difference, by the way, if each bounce was a third, or half or seven-eighths as high as the previous one. The only thing that would change would be the time in which an infinite number of bounces took place. It might turn out to be one second, or seven seconds, or ten seconds, instead of four.

In real life one could not get a ball to bounce an infinite number of times. No ball is a perfect sphere, and no floor is perfectly smooth. Both surfaces would appear irregular when examined with a powerful microscope. As a result, friction would cause the motion to cease after some finite number of bounces. In fact, when I tried bouncing a Ping-Pong ball on my kitchen floor, it bounced about twenty times and then rolled under the dishwasher. An infinite number of bounces does seem to be possible in principle, however. Under such circumstances, no matter how many times the ball has bounced, it will bounce an infinite number of times in the future. As though this were not strange enough, it appears to be quite difficult to say precisely what the ball is doing after exactly four seconds have elapsed. Is it traveling upward at some infinitesimal velocity? Downward? Or has it come to a halt?

It appears that it is not so easy to answer questions about infinity when they are placed in a real-world setting. It should be noted, incidentally, that Cantor's concept of transfinite numbers is of no help to someone who wants to refute Zeno or to answer questions about bouncing balls. All that Cantor's theory tells us is that the series ½, ¼, ⅛, 1/16 . . . can be placed into a one-to-one correspondence with the positive integers 1, 2, 3, 4 . . . and that consequently both infinite sets are the same size. It tells us nothing about the behavior of real-world objects. Though Cantor taught us that it was legitimate to talk about infinite numbers, his work did not make them seem less strange. If anything, his theories had the opposite effect. Perhaps we should consider ourselves fortunate that we don't often encounter infinities in real life.

THE PASSAGE OF TIME

The American philosopher William James created his own version of Zeno's paradox, "The Dichotomy." Fourteen minutes can never pass, according to James. First, it is necessary that seven minutes go by. Before the seven, three and a half minutes must pass, and before that a minute and three-quarters, and so on. In a similar manner, one can "prove" that Zeno should never have died. After all, before his life could be completed, he had to live through the first half, and then half of what remained, and then half of that, and so forth. However, that doesn't seem to be the way things worked out. During the latter part of his life Zeno became involved in political activities in his native city, Elea. According to stories that circulated in late antiquity, the tyrant who ruled the city had Zeno tortured to death, and Zeno endured his fate heroically. One can't be entirely sure that the stories are accurate—they don't agree as to the details of Zeno's death, and the tyrant is given a number of different names.

MEANWHILE, BACK
AT THE CELESTIAL HOTEL . . .

The Babe (George Herman was generally known by the name "Babe" during his lifetime) was having an argument with the Celestial Yankees' shortstop. The Babe maintained that if he turned on a light in one of the rooms of the Celestial Hotel on Monday, turned it off on Tuesday, turned it on again on Wednesday, and repeated the same pattern for all eternity, its final state after an infinite number of switches would be off. The shortstop maintained that if the light was turned on, then off, then on—and then off and on for all eternity, then its final state would be on. Neither seemed to be able to understand the other's position.*

The discussion was overheard by Yogi, who happened to be walking down the hotel corridor. He had heard them argue in this manner many times before and knew that they would do it an infinite number of times in the future. He listened for a few moments, just long enough to convince himself that they were at it once again. Having no desire to become involved in the argument, he turned around and walked the other way, muttering, "It's déjà vu all over again," as he walked away.

Then, as the Babe walked down the hotel corridor, he had a sudden, terrifying thought. He was going to play an infinite number of games. Yet only a finite number of plays were possible in baseball. If someone hit a

---

* It may seem I wrote this passage to amuse, but that is not the case. The question of the final state of the lightbulb is one that has appeared in philosophical discussions about the nature of infinity. It is a problem that seems to have no solution.

fly ball, for example, there were only six possibilities: a single, a double, a triple, a home run, a foul ball, or an out. If a runner tried to steal second, the number was even smaller. Either he was safe or he was out. A finite number of possibilities could be combined in only a finite number of ways. Hence he was destined to play the same games over and over and over again an infinite number of times.

Suddenly the Babe's frown changed into a smile. No, that wasn't true, he decided. In any given inning, a team could get any number of hits before there were three outs. In fact, there were an infinite number of possibilities. Then he found himself becoming puzzled again. Did this mean that some games would have innings of infinite length?

2

INFINITE TIME

THE BRITISH MATHEMATICIAN Charles Lutwidge Dodgson—better known to us by his pen name Lewis Carroll—once parodied Zeno in a tale called "What the Tortoise Said to Achilles." In this story, which he published in the British philosophical journal *Mind*, we find Achilles and the Tortoise discussing a simple theorem in geometry, the First Proposition of the Greek mathematician Euclid. Carroll states the theorem, which deals with a triangle, as follows:

> **(A)** Things that are equal to the same are equal to each other.
> **(B)** The two sides of this Triangle are things that are equal
>   to the same.
> **(Z)** The two sides of this Triangle are equal to each other.

It is obvious to Achilles that if A and B are true, then the truth of Z logically follows. This is disputed by the Tortoise, who says that he will accept A and B, but that he is not so sure about Z. To satisfy him, Achilles introduces another premise:

> **(C)** If A and B are true, Z must be true.

The Tortoise must now surely accept the conclusion Z, Achilles says. Not so, says the obstinate turtle. He accepts A, B, and C, but not Z. Consequently, Achilles adds yet another premise:

(D) If A and B and C are true, then Z must be true.

As you've probably guessed, this kind of thing can go on forever. At the end of the story, over a thousand premises have been written down, and the Tortoise comments that there are "several millions yet to come."

The tale isn't quite as witty as Carroll obviously intended it to be. Moreover, it contains some very bad puns—at one point, the Tortoise is renamed "Taught-Us" and Achilles "A Kill-Ease." Yet, the story does offer a good illustration of the impossibility of an infinite regress. The Tortoise is guilty of understatement when he says there are "several millions yet to come." Obviously, however many premises Achilles adds, the Tortoise can always ask for one more. This process can go on forever.

Naturally, the impossibility of an infinite regress often arises in philosophical arguments, especially in arguments that purport to demonstrate the existence of God. Aristotle "proves" the existence of God by the argument of a First Mover. Everything that is in motion, Aristotle says, must be moved by something else. Body A is moved by body B, which is moved by body C, and so on. But this series cannot go on forever. There must be some First Mover, which is God.

Aristotle isn't saying that God initially set the universe in motion. In his philosophy, motive force must be applied continuously, or motion will stop. The reason that a stone continues to move after it has left the thrower's hand, for example, has nothing to do with momentum; that concept was unknown in Aristotle's time. A stone was thought to be propelled along by the motions of the air through which it passed. When these motions were no longer sufficient to continue pushing it along, the stone would drop perpendicularly to the ground.* Similarly, if the First Mover would somehow cease to exist, then all motion in the world would also cease.

The argument was eventually to become part of Roman Catholic theology. In the thirteenth century, St. Thomas Aquinas wrote a monumental book entitled *Summa Theologica*, in which he labored to reconcile the

---

* Of course, a thrown object doesn't behave anything like this at all. It is clear that Aristotle never had to catch a fly ball.

philosophy of Aristotle with Church doctrine. It is not surprising to see Thomas repeat the argument of the Unmoved Mover. In fact, Thomas gives five proofs of the existence of God, two of which are variations of the Unmoved Mover idea. In the better known of the two, he invokes the idea of a First Cause. Every event must have a cause, Aquinas says. But this series cannot go on endlessly. There must therefore be a First Cause. In yet another argument, he says there must be a source of all necessity.

Such reasoning seems quite unconvincing to the modern scientific mind. Any individual scientist may or may not believe in God. Whether he does or not, he would not be likely to accept these, or similar, lines of argument. Aristotle believed that objects would eventually come to rest if there was not some force moving them. We know today that motion is the natural state of any body. One of Sir Isaac Newton's laws of motion states that any body that is in motion will continue to move in a straight line unless some external force acts upon it. Similarly, the idea of a First Cause sounds somewhat fishy in light of the modern theory of quantum mechanics. According to the most commonly accepted interpretation of quantum mechanics, individual subatomic particles can behave in unpredictable ways and there are numerous random, uncaused events.*

Other arguments involving infinite regress are not so easy to dispose of. In his book *Critique of Pure Reason*, the eighteenth-century German philosopher Immanuel Kant argued that it was absurd to suppose that time was infinite. If an infinite quantity of time had elapsed before the present, then an infinite number of events must have taken place, which was impossible.

Kant was not trying to prove that time was finite. In fact, he presented another argument that seemed to lead to the opposite conclusion; he believed he could show that the idea of finite time also led to a contradiction. Time could not have a beginning, he said, because one could then ask what had happened before this point. It could not have an end because one could then ask what happened after that. Kant was trying to

---

* We'll return to quantum mechanics in a subsequent chapter.

prove that, since time could be neither infinite or finite, it was not a property of the external world. On the contrary, it had to be something that was innate within the human mind, not a characteristic of the external world.

It is not my intention to expound German idealistic philosophy, however, only to give some examples of arguments that make use of infinite regress, especially as it applies to questions concerning the nature of time. As we shall see, the question of the infinity or finiteness of time is one that has been debated since ancient times. Even today, scientists do not really know whether time is infinite or finite. Most of them believe that time began in the big bang in which our universe was created. Whether or not time will eventually come to an end is not known.

Anyone who begins to think about the nature of time is immediately confronted by questions about infinity. Infinity and eternity, after all, are closely related concepts. Thus I think it would be of interest to backtrack and see how the problem of time was dealt with during various periods of history, beginning with the ancient world. One of the things that makes such an endeavor interesting is that certain very old ideas are very much like modern ones. For example, we will find that some of the ideas about time that were expounded in ancient Greece bear an uncanny resemblance to concepts that have arisen in the context of modern cosmology.

CIRCULAR TIME

Aristotle maintained that the world had always existed. At first glance, this is a surprising statement. After all, not only does he appeal to the impossibility of an infinite regress in his Unmoved Mover argument, Aristotle also denied that infinite quantities could really exist. Only potential infinities exist, he said.

The set of positive integers provides a good example of what Aristotle meant. In Aristotle's terminology, the numbers 1, 2, 3, . . . are potentially infinite because, although you can go on counting as long as you want, you will never get all the way to infinity. For any number that you can imagine, it is always possible to find a larger number. If one has the number

568,922,777,983,416,715,334,065,784,930,891,419,851,002,713,842
it is only necessary to add 1 to obtain the larger number 568,922,777,
983,416,715,334,065,784,930,891,419,851,002,713,843. For that matter, one could add 13 to obtain 568,922,777,983,416,715,334,065,784,
930,891,419,851,002,713,855, or add 100 trillion, to get 568,922,777,
983,416,715,334,065,784,930,891,519,851,002,713,842. Though very
large, these numbers are far from being infinite. What Aristotle called an
actual infinity is never encountered.

How, then, could Aristotle conclude that past time was infinite? In reality, he didn't. According to Aristotle, time was a circle. He accepted a
belief that was quite common in ancient times, that of cosmic cycles. We
think of time as something that stretches in a straight line from the past
into the future. Yet it is not really more natural to think of time in this way
than it is to regard it as circular. After all, most natural phenomena are
cyclic. We all experience the rhythms of night and day, the phases of the
moon, and the progressions of the seasons.

Indeed, the idea of cyclical time has arisen in many different civiliza-
tions. The Hindus conceive of cosmic cycles of vast length and believe
that the world is periodically destroyed and re-created. During the Vedic
period (about 1500 to 600 B.C.), Indian sages elaborated upon this and
conceived of cycles within cycles. The smallest was an age, about 360
human years. The longest, which were on the order of 300 trillion years,
corresponded to the lives of the gods.

The ancient Chinese believed in a cyclical interplay between the op-
posing cosmic principles of yin and yang, and calculated a cycle of
23,639 years. In the New World, the Aztecs and Mayas believed in cyclical time and cyclical catastrophes. They were somewhat more moderate
in their conception of time than the Hindus, however. According to Aztec
belief, the world was in danger of being destroyed every fifty-two years.
We also find the idea of world cycles in Norse mythology. According to
Norse myth, the earth and the heavens would be destroyed in a final battle between the gods and the giants. The world would be created anew,
with new gods and a new human race. Presumably this was a process that
would be repeated over and over and over again.

THE GREAT YEAR

In ancient Greece, it was commonly believed that a cycle existed known as the great year. When, after thousands of years, the sun, the moon, and the five known planets regained a certain original configuration, time would have completed its cycle. In every cycle, there would be a "great winter," during which the world would be inundated with rain and floods, and a "great summer," during which it would be destroyed by fire.

Many believed that in each cycle the events of human history would be repeated exactly. There would be another Trojan War, another Athens, even another Socrates and another drinking of the hemlock. This was one of the doctrines propounded by the mystical brotherhood called the Pythagoreans, after its founder, Pythagoras. It is also encountered in Plato's dialogue *Parmenides*, when Parmenides states that something that is becoming older must simultaneously be getting younger. The idea is that the thing that is becoming older is simultaneously moving away and getting nearer to its beginning in circular time. Parmenides does not elaborate upon this idea; presumably he expected that his listeners would understand what he meant.

The concept of cyclical time is also encountered in *Problems*, a book traditionally attributed to Aristotle. At one point the book speaks of living both before and after the Trojan War. The idea is that if time is circular, then the Trojan War is in both the past and the future. This probably does not represent Aristotle's true opinion, however, as modern scholars believe *Problems* was compiled long after his death.

If time is a circle, there is no infinity of past events. Passing through an entire cycle of time or great year would be like making a journey around the earth; you would eventually come back to the same place. In fact, it isn't even necessary to speak of an infinity of cycles. On the contrary, the very same cycle can be repeated over and over and over again.

Aristotle did not believe that past events would be repeated exactly, or that the same individuals who were alive in his time would live again in the future. He does, however, make reference to the concept of the great year and such events as the great winter. Furthermore, he believed that human events would follow certain cyclic patterns.

To the modern mind, it may seem odd to conceive of time as a circle if past events do not repeat themselves. But perhaps we should remember that we do sometimes speak of time in this manner. We often talk of getting up at the "same time" every morning, or of going to work, or eating dinner or going to bed at some particular time. Aristotle is making use of the same concept, but on a much grander scale. In his view, it is the "same time" when certain particular heavenly motions repeat themselves during successive great years.

THE STOICS

In spite of Aristotle's skepticism, the idea that events did repeat themselves exactly became extraordinarily influential. For that idea was one of the doctrines associated with Stoicism. Though the earliest Stoics were Greek, Stoicism became a popular doctrine among the Romans. In fact, it acquired some of the characteristics of a religion. There is nothing particularly surprising about this, by the way. In ancient times, the distinction between philosophy and religion was not nearly so great as it is today. Numerous religious overtones are found in the writings of Plato, for example, and in Roman times, people often looked to philosophy when they sought a guide to the conduct of life.

Stoicism was founded by Zeno of Citium (who should not be confused with Zeno of Elea) around 300 B.C. Little of Zeno's writing survives today; we possess only a few fragments. Nevertheless, it is possible to obtain a fairly clear picture of his doctrines. Interestingly, it seems Zeno didn't intend to propound an original philosophy. Rather, he was more concerned with collecting ideas that represented the "wisdom of the ages." In doing so, however, he put together a philosophy that endured for centuries. Nearly five hundred years after Zeno died, eminent Romans were still propounding Stoic ideas.

According to Zeno the same events were destined to repeat themselves in endlessly recurring cycles. At the end of each cycle, the entire cosmos would be destroyed in a great conflagration, and then it would be born anew. Initially, there would be nothing but fire. Gradually, the fire would condense into air, and then into water. Earth would form from some of the water, and a new world would be created, to endure until everything was eventually transformed into fire again. Though this sounds like an odd series of transformations to the modern mind, it would have seemed perfectly logical in Zeno's time, when it was thought that the four elements— earth, air, fire, and water—could be transformed into one another.

Stoicism remained in vogue for centuries, so it is natural that some of the doctrines associated with the philosophy should have changed in time. In particular, greater emphasis was placed on ethical and religious ideas in later Stoicism. However, one fundamental tenet remained constant: human beings were powerless to alter the course of events and were destined to live the same lives over and over again. But there was one sense in which they were free: they could cultivate inner virtue and harmony. Virtue was the only good; worldly matters like health and material possessions were of no importance.

Stoicism sounds like an austere philosophy. Today it seems amazing that a doctrine teaching that all actions are pointless was so widely accepted. Yet Stoicism had a great appeal in its day, and it endured for an amazingly long time. It became especially popular among Romans of all classes. One noted Stoic philosopher, Epicetus, was a slave; another, Marcus Aurelius, was an emperor.

Among those who popularized the Stoic doctrine in Rome was the politician and prosecuting attorney Cicero.* Cicero was an opponent of Julius Caesar, and though he didn't participate in Caesar's assassination, he sided with those who thought it necessary. An enemy of Marcus Antonius (who is better known to us as Marc Antony), Cicero was executed by the Emperor Augustus after the latter assumed power. In the midst of an eventful political life, Cicero found the time to write numerous books, including one called *The Nature of the Gods*, in which he expounded on three rival philosophies—one of which was Stoicism.

Strictly speaking, Cicero was not a member of the Stoic school of philosophers, but to include him among the Stoics is not to stray far from

* Those who look up Cicero in an encyclopedia will read that he was "Rome's greatest orator" or something of the sort. I prefer to describe him in a somewhat more concrete way. The two descriptions aren't that different; the main uses of oratory were in politics and in the courts of law.

the truth. He was drawn to the school's doctrines, and many of his own writings have a Stoic tone. Of the three philosophies Cicero discusses in his book, it is Stoicism that is described most sympathetically.

Other eminent Stoics included Seneca, a Roman philosopher, tragedian, and statesman, and Marcus Aurelius. Seneca was tutor to the emperor Nero and was influential in the political affairs of the day. Eventually, he fell out of favor and Nero ordered him to commit suicide. Seneca, incidentally, seems to have been one of those philosophers who was unable to follow the dictates of his own doctrines. As a Stoic, he was supposedly indifferent to wealth, yet he acquired an enormous fortune. Some of it was obtained by lending money in the Roman province of Britain at excessively high rates, and this may have been one of the factors that caused the Britons to revolt against Roman rule.*

Marcus Aurelius became emperor in A.D. 161. Today he is known mainly for his *Meditations*, a book of philosophical musings. These meditations, which were often written in the midst of military campaigns, were apparently never intended for publication. Aurelius seems to have written them only as an exercise in self-consolation. He would have preferred a quiet life, but the duties of his office prevented this. During his reign, a whole series of disasters befell Rome; there were plagues, earthquakes, insurrections, and long wars. Aurelius was one of a series of Roman emperors who persecuted the Christians. He seems to have done this out of a sense of duty. The Christians rejected the state religion, which Aurelius thought of as a political necessity.

THE STOIC CONCEPTION OF TIME

The Stoics believed that time was circular and finite, but, unlike Aristotle, they had no horror of infinity. According to Stoic cosmology, the visible cosmos—the earth and the heavens—existed within a void of infinite

---

* The immediate cause of the revolt was the flogging of Queen Boadicea and the rape of her daughters by Roman soldiers. Afterward, Boadicea and her Celtic warriors nearly drove the Romans out of Britain.

extent. Beyond the stars and the planets, there was only empty space, a space without end or boundaries.

Aristotle had refused to concede there was anything outside the cosmos. He rejected the idea of an external void. The Stoics didn't agree. To show that a void had to exist, they used an old argument that seems to have originated with the Pythagorean philosopher Archytas of Tarentum. The argument runs something like this: Suppose someone stands at the edge of the cosmos and thrusts his arm outward. What will happen? He will extend it into the void. Now imagine him standing a little farther out and extending his arm again. Obviously this process can be continued indefinitely. This, the Stoics said, proves that the void is infinite. The use of this argument, incidentally, shows that the Stoics conceived of infinity in a somewhat different manner than Aristotle did. To Aristotle, such reasoning would have shown only that such a void—if one could indeed exist—was *potentially* infinite. To the Stoics, on the other hand, infinity was something very real.

The Stoics conceived of the cosmos as an object inside an infinite sea of nothingness. Inside the cosmos, events repeated themselves endlessly, and the only thing that was free was the human will. Under such circumstances, could one reasonably attempt to do anything but be "stoic"? The Stoic philosophers didn't think so. If you could not alter the course your life was destined to take, it was best to concentrate on cultivating your inner virtue.

LINEAR TIME

As I remarked previously, we habitually think of time not as the ancient Greeks did, but as something that stretches in a straight line into the past and the future, possibly into the infinite past and infinite future. This concept of linear time is a consequence of our Judeo-Christian heritage.

In ancient Judaism, there was no place for cosmic cycles. The world was created at a particular point in time. God created the world from a formless void. After spending six days on the task, He rested. Furthermore, we are told that certain events happen only once; they will not be repeated. For example, the Book of Genesis tells us that God promised Noah there would never be another Deluge and that He created the rainbow as a token of this promise. Similarly, the Exodus was something that happened once. The Jews were not destined to flee from Egypt again and again and again.

Traces of the idea of circular time are here and there in the Old Testament. We read in Ecclesiastes that "the thing that hath been, it is that which shall be; and that which is done is that which shall be done: and there is no new thing under the sun." Such references are few, however, and the existence of a few isolated passages of this character doesn't show that the idea of circular time was common in ancient Jewish culture.

By its very nature, Christianity emphasized the concept of linear time to an even greater degree. Christian doctrine, after all, centers around the suffering, death, and resurrection of Jesus. If it were implied that this happened over and over again in succeeding cosmic cycles, the entire meaning of the Redemption would be lost. Thus in the letters of St. Paul, which make up part of the New Testament, we find Paul insisting on more than one occasion that Jesus suffered *once* for humanity's sins.

Approximately three hundred and fifty years later, St. Augustine felt it was still necessary to emphasize that "once Christ died for our sins." Augustine, who had become bishop of the city of Hippo in North Africa in A.D. 396, began writing his great book, *City of God*, in 412. In 410 Rome had been sacked by the Visigoth king Alaric I, and many had attributed this disaster to the abandonment of the old pagan gods in favor of Christianity. Augustine felt that this charge had to be answered, and set out to write a reply. But his book was not finished until 427, and as Augustine wrote, it developed into something much greater than the tract he had originally intended. Along the way, he attacked the "deceiving and deceived sages" who continued to advocate the idea of cyclical time, and astrology as well. Not only was the cyclical conception of time a foolish doctrine, Augustine said, it was also impious. Echoing St. Paul, he emphasized that "once Christ died for our sins; and, rising from the dead, He dieth no more."

Augustine was well aware of a strong connection between the concept of a great year and astrological doctrine. For if cycles of events are repeated endlessly in accordance with the movements of the planets, it would follow that it should be possible to foretell future events by studying the heavens. But astrology could not be true, Augustine said, arguing that twins, who had the same horoscope, often had different fates. Augustine was bothered by the determinism associated with astrology. This may have been one of the factors that caused him to attack the Stoic conception of time, where determinism and cyclical time go together.

Although the idea of cyclical time was opposed by the Church, belief in the doctrine persisted for many centuries. In 1277, when Etienne Tempier, the bishop of Paris, cited some 219 opinions to be condemned as heretical, the doctrine of cyclical time was sixth on the list. And, many years later, during the Renaissance, the idea again became a topic of intense discussion. It appears that even in Western culture, where religious doctrines have led to an insistence upon the idea that time is linear, the idea of unending cosmic cycles has always had a certain appeal.

TIME AND ETERNITY

Accepting the Judeo-Christian doctrine of a creation at a particular point in time doesn't eliminate the problems associated with the concept of infinite linear time. Not only do the paradoxes that were cited by Kant remain, they initially seem to get a little worse. If the world was created at a particular point in time, does this mean that God allowed an eternity to pass before He began the Creation? And why didn't He create the world sooner than he did?

The "deceiving and deceived sages" who advocated the idea of circular time may have posed such questions to St. Augustine, for in his *Confessions* he speaks of those who asked, "What was God doing before He made the heaven and the earth?" It is sometimes said that Augustine answered, "Preparing hell for those who ask such questions." In reality, he wrote that he would not give so frivolous an answer as that, and he made a serious attempt to find a reply.

Augustine was steeped in Greek philosophy. However, the Greeks were not much help. These philosophers had generally viewed matter as something that had always existed. This idea is even found in the works of Plato, which do not advocate the idea of cyclical time. To be sure, Plato speaks of a creation in his dialogue *Timaeus*, but Plato's demiurge is not a Creator; he simply gives form to matter that already existed.

Augustine's solution to the problem therefore had to be an original one, and it was. He concluded that time simply did not exist before the Creation. Time and the world came into being together. God's eternity was not a kind of time; on the contrary, God remained eternally outside of time. To ask what God was doing before the Creation was to pose a meaningless question. There was no before.

The questions that Augustine was asked and the answers he gave bear a remarkable resemblance to the questions and answers we sometimes hear today. Fundamentalist Christians will ask, "And what was happening before the big bang?" The answer scientists give is basically the same as Augustine's: there was no "before." In the big bang, time, space, and matter were all created together.

But one should not imagine that Augustine solved all the problems associated with the idea of unending linear time. He certainly didn't seem to think so. At one point, he asks somewhat rhetorically, "What, then, is time?" He then answers, "If no one asks of me, I know; if I wish to explain to him who asks, I know not." At another point, he says, "I confess to Thee, O Lord, that I am yet ignorant what time is."

The topic of the nature of time is a fascinating one. But we would be led too far astray if I pursued it at any great length.* However, I will return to the question of the infinity or finiteness of linear time in subsequent chapters. As we shall see, scientists still do not know whether time is finite or infinite.

IVAN KARAMAZOV AND THE DEVIL

There is a passage in Dostoevsky's late nineteenth-century novel *The Brothers Karamazov* where Ivan Karamazov hallucinates that the Devil

---

* I refer the interested reader to my book *Time's Arrows* (Simon & Schuster, 1985).

has come to visit him. The Devil, dressed as a somewhat seedy-looking Russian gentleman, professes to be just as much of an agnostic as Ivan is. But he is a pleasant fellow and an interesting conversationalist. During the course of his conversation with Ivan, the Devil expounds on the ancient theory of cosmic cycles. The earth has been destroyed and re-created "as many as a billion times," the Devil says. Furthermore, the process will repeat itself "infinitely and always in the same way."

Interestingly, the Stoic idea of cosmic cycles now appears within the context of linear time. As a Christian, Dostoevsky accepts the idea that there has been a creation. Thus he doesn't have the Devil say that the cycle has repeated itself an infinite number of times in the past. The number of past cycles has been "as many as a billion." On the other hand, Christian doctrine doesn't prevent him from imagining an eternal future. So these cycles can repeat themselves an infinite number of times.

The idea is occasionally encountered in contemporary literature also. In *The Unbearable Lightness of Being*, the Czech novelist Milan Kundera speaks of the "mad myth" of eternal return, and says, "If every second of our lives returns an infinite number of times, we are nailed to eternity as Jesus Christ was nailed to the cross. It is a terrifying prospect."

When he speaks of eternal return, Kundera is making reference to an idea proposed by the nineteenth-century German philosopher Friedrich Nietzsche, who was a contemporary of Dostoevsky. Nietzsche didn't simply restate the Stoic conception of time; rather, he argued that the idea of endless cycles was quite plausible. If the universe contained only a finite number of atoms, Nietzsche said, then these atoms could have only a finite number of different configurations. But, Nietzsche went on, time was infinite. Any given configuration of atoms on our earth would be repeated innumerable times in the future. The events happening today would take place again and again and again.

Nietzsche was a classical scholar, and his doctrine of eternal recurrence grew out of his readings of ancient philosophers. Nevertheless, there is a difference between his theory and theirs. Nietzsche does not think of time as circular. He assumes that it is linear and infinite. Yet he reaches the same conclusion reached by his ancient predecessors: events will repeat themselves endlessly.

Nietzsche's theory never attracted the attention of scientists, and today some of the questions it raises seem moot. As we will see later on, Einstein's general theory of relativity implies that if the universe is infinite in extent, then time must be infinite, too. And if the universe is finite, then time is finite also; it will eventually come to an end. If Einstein's theory is correct—and we have every reason to believe it is—then Nietzsche's assumption of infinite time in a finite universe is wrong.

3

INFINITE WORLDS

THE QUESTION of whether the universe was finite or infinite was often debated by medieval scholars. Their conception of an infinite universe bears little resemblance to modern ones. Medieval philosophers generally followed Aristotle in assuming that the earth was surrounded by a series of moving celestial spheres. The innermost sphere contained the moon; other spheres were assigned to Mercury, Venus, the sun, Mars, Jupiter, and Saturn, in that order. Outside of Saturn's sphere was the sphere of the fixed stars, which rotated rapidly around the earth, completing a revolution every twenty-four hours. Sometimes additional spheres were added for theological or philosophical reasons, so there might be anywhere between eight and eleven heavenly orbs in all.

The fundamental question was what, if anything, lay beyond the outmost sphere. Some stuck by Aristotle, maintaining that nothing existed beyond the boundaries of the cosmos. There was no matter, no space, no time. According to Aristotle, the very notion of a void beyond the outermost sphere was self-contradictory. Some medieval philosophers disputed this conclusion and maintained there was an infinite void.

Although this conception of the universe resembled that of the Stoics, the reasons for postulating an infinite void were generally different. If God was infinite, some philosophers said, there must be an infinite space in which He was omnipresent. According to an often-quoted metaphor, drawn from the twelfth-century work *Book of the XXIV Philosophers*, "God is an infinite sphere whose center is everywhere and circumference nowhere."

The discourse about the universe arose out of Roman Catholic theological debate. Many of the arguments that were developed, though ingenious, are generally of interest only to scholars. After all, we know today that medieval cosmology was wrong. The earth is not surrounded by a series of heavenly spheres, and if the universe is infinite, it is made up not of an empty, immeasurable space, but of infinite numbers of galaxies and stars.

THE DEVELOPMENT OF ASTRONOMY

The idea that the cosmos was made up of a series of heavenly spheres with the earth at the center remained unchallenged for centuries. It was backed by the authority of both Aristotle and the Bible, which spoke of a fixed, unmoving earth. Furthermore, astronomical knowledge developed slowly during the Middle Ages, and astronomy generally had little connection with cosmology. Astronomy was a highly technical mathematical subject. The main impetus for its study was the desire, during the late Middle Ages and the Renaissance, to cast more accurate horoscopes. Cosmology, on the other hand, was the domain of theologians and natural philosophers, who generally lacked the mathematical training that would have allowed them to peruse astronomical treatises.

Even if the philosophers had been more familiar with the astronomical knowledge of the day, it might not have made much difference, for there was no single, agreed-upon astronomical theory. By the sixteenth century there were more than a dozen. To be sure, these different systems shared certain assumptions: the earth was the center of the cosmos, and the sun, moon, and planets moved around it. However, each system analyzed the details of these motions differently.

If one assumes that the earth is immobile, two different kinds of motions are observed in the sky. The fact that the earth rotates on its axis causes the sun, moon, and planets to rise and set. The revolution of the earth around the sun produces another, much slower, motion. In summer, when the earth is on one side of the sun, the sun appears against a certain background of stars. In winter, the stars that appear behind the sun are different ones. Of course one can't see these stars in daytime, but it is not difficult to calculate the sun's position. This is reflected in our astrological signs. A person born between March 21 and April 19 is said to have been born when the sun was in the constellation Aries, while someone with a birth date between December 22 and January 19 is said to have been born when the sun was in Capricorn.

In the case of the various planets, we can observe this directly. The planet Mars, for example, sometimes appears to be in the midst of one group of stars, and sometimes in the midst of another. It should be apparent that, since both Earth and Mars revolve around the sun, their relative positions will change. As they do, Mars will appear to move from one place in the sky to another.

If one assumes that the sun, moon, and planets revolve around a motionless Earth, these apparent motions against the background of the fixed stars become quite complicated. Even the motion of the sun is not easy to describe. The sun, which is closer to Earth in summer than in winter, appears to travel at different speeds during different seasons. It cannot be assumed that it has a uniform circular motion, because then theory will not conform to observation. Adjustments must be made. In the case of a planet, the apparent motions are even more complex. The relative positions of Earth and Mars, for example, depend upon Mars' orbital motions as well as those of Earth. As a result, not only does Mars seem to move across the field of stars at different velocities at different points in its orbit, at times it seems to reverse course and move in a retrograde manner; that is, in a direction opposite to that in which it ordinarily moves. It then changes course a second time and resumes its progress in its original direction.

It is not difficult to see why this should happen. Earth, which has an orbit inside that of Mars, moves with a greater velocity. Thus there will be times when Earth "passes" Mars in a manner similar to that in which a runner on an inside track may pass one on an outer track. Under such circumstances, the background of stars behind Mars appears to shift rapidly.

Late medieval and Renaissance astronomers used a number of mathematical techniques to account for the apparent motions of the planets. All of the systems made use of epicycles. However, epicycles alone could not account for the observed motions of the planets, so astronomers put epicycles upon epicycles. Sometimes they assumed that a planet revolved, not around Earth, but around a point in space near Earth. Alternatively they might make the assumption that planets moved at different speeds at different points in their orbits. There was no general agreement as to exactly how planetary motions should be described. Different astronomers used different combinations of techniques to work out the details. Where one might add an epicycle to a system that already made use of many epicycles, another would use a different mathematical technique. Interestingly, the astronomers generally paid little attention to the idea of celestial spheres. This is another example of the division between astronomy and philosophy in this era.

By the end of the sixteenth century, astronomical systems had become quite complicated. Yet none of them quite worked. They all failed to agree with astronomical observations to some extent. Thus, when the Polish astronomer Copernicus was asked to advise the Church about calendar reform early in the sixteenth century, he suggested that the project be postponed. He felt that astronomical knowledge was not complete enough to enable a truly accurate calendar to be constructed. This is certainly a revealing statement. After all, to create an accurate calendar, it is not necessary to know anything about the motions of any of the planets. It is only necessary to understand the apparent motions of the sun and moon. In effect, Copernicus was saying that these could not yet be accurately calculated.

THE COPERNICAN REVOLUTION

Copernicus waited until the end of his life to publish *De Revolutionibus Orbium Coelestium* (On the Revolutions of the Heavenly Orbs), the book in which he expounded the theory that the sun, and not the earth, was the center of the solar system. It was published in 1543, the year of his death. Indeed, he is supposed to have received the first printed copy upon his deathbed.

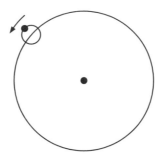

Figure 1. The epicycle is one of the simpler devices used in pre-Copernican astronomy. The black dot represents the earth. The larger circle is the deferent, or primary orbit of a planet around the earth. The planet is not located exactly on the deferent, but on an epicycle. The planet moves along the epicycle while the center of the epicycle moves along the deferent. If the two motions are in opposite directions (one clockwise, one counterclockwise), the planet, as seen from the earth, will sometimes seem to reverse direction. Pre-Copernican astronomers often went further and placed epicycles upon epicycles, and even epicycles upon epicycles upon epicycles.

The reception accorded to Copernicus's book was quite different from that given to Galileo's writings about the Copernican system ninety years later. In 1633, Galileo was tried by the Inquisition on charges of heresy. Threatened with torture, he was forced to recant his advocacy of heliocentric astronomy. Anyone familiar with this story might imagine that the publication of *De Revolutionibus* would have also aroused strong opposition, but this was not the case, for two reasons. Copernicus's theory was so new that no one had yet considered what its theological implications might be. Second, his book was highly technical in nature and virtually unreadable to anyone but astronomers. It quickly became known in the astronomical community, but few others were aware of its existence.

Astronomers were quick to appreciate the importance of Copernicus's work. Few of them were willing to accept the idea that the earth revolved around the sun. They realized, however, that the theoretical tools that Copernicus had developed were useful for calculating the positions of the planets and were willing to accept the idea of a moving Earth as a useful mathematical fiction.

Actually, Copernicus's system was not appreciably simpler than those previously in use. Copernicus had made the seemingly natural assumption that planetary orbits were circular. In fact, they are not. Planets move in elongated orbits, which have the form of geometrical figures known as ellipses, and their orbital velocities vary according to their distance from the sun. In order to account for the fact that the sun appeared to move faster in winter than it did in summer, Copernicus was forced to assume that the earth revolved around a point near to the sun, not around the sun itself. And, though his system eliminated the necessity of assuming that the planets sometimes changed direction and moved in a retrograde manner, Copernicus still had to make use of epicycles and other techniques used by his predecessors in order to get his theory to agree with astronomical observations.

One of the reasons that Copernicus's contemporaries were reluctant to accept the idea that the earth might revolve around the sun was that no one had ever observed any shifts in the apparent positions of the stars. This was exactly what one would expect if the earth remained fixed in the center of the cosmos.

Modern measurements place Earth at a distance of about 93 million miles from the sun. When it travels from one side of its orbit to the other, it is displaced by twice this distance, or 186 million miles. Earth's orbit is almost circular, and the diameter of a circle is twice its radius. The stars should therefore look somewhat different in winter than they do in summer. After all, a tree does not look the same when viewed from a position a little to the right as it does when viewed from the left. When viewed from different vantage points, the positions of the stars should similarly seem to be a little different.

And this is exactly what happens. But the stars are so far away that this effect, called stellar parallax, is very small. It was not observed until 1838, well after the invention of the astronomical telescope. There was simply no way that stellar parallax could have been seen by naked-eye observers of the sixteenth century.

Copernicus realized that stellar parallax should exist, but he was unable to observe it. He attempted to solve the difficulty by suggesting that the stars were so far away that the effect was too small to be seen. This was a new idea. It had always been assumed that stars were not so much farther from Earth than Saturn, the most distant known planet. For example, during the ninth century, the Arab astronomer al-Fargani estimated that the sphere of stars was about 75 million miles from Earth. In his day, and in Copernicus's time as well, this sounded like a huge number. No one imagined that the distance to the nearest star would eventually turn out to be more than 300,000 times greater.

OTHER WORLDS

In Copernicus's time, there was no way of calculating how far away any of the stars might be, or of determining whether they lay on a sphere—or whether they were positioned at varying distances from the sun. But one thing was obvious. If the Copernican theory was indeed correct, then the stars were not tiny points of light. In order to be seen at such great distances, they must be very large and bright, perhaps as bright as the sun itself.

The first person to exploit this idea seems to have been Thomas Digges, an Englishman who published a popular account of Copernicus's theory in 1576. Realizing that the stars did not have to be positioned on a sphere, Digges described a universe in which the stars were other suns. The earth and the sun were at the center, and the stars were scattered without limit in every direction, in an infinite space that surrounded the solar system.

Digges's ideas were somewhat self-contradictory. There can only be a "center" in a finite space that has definite boundaries. If space is infinite, every point is as much a center as any other. But if Digges failed to see this, it is of little importance. He did understand the more important point that, if Copernican theory was correct, then the universe was either infinite or immeasurably large.

The idea of an infinite universe is generally associated not with Digges, but with the Italian philosopher Giordano Bruno, who made the idea of infinite worlds a central part of his philosophy. Bruno, born in 1548, became one of the best known scholars of his day. He was not afraid to expound ideas that were unconventional or even heretical. For this, he paid a heavy price—he was arrested for heresy in 1591, tried twice, and finally burned at the stake in the year 1600.

Bruno was rebellious and argumentative by nature. In 1565 he joined the Dominican order, which seems not to have been a wise decision, for his rebelliousness and insubordination to the monastic authorities led eventually to charges of heresy, filed against him in 1576. He fled the monastery and took up what was to be the life of a wandering scholar. During that year, Bruno lived for short periods of time in the Italian cities of Genoa, Turin, Savona, and Noli. From there, he went to Venice, then to Padua and Milan. Finally he left Italy in 1578 and took himself to France, where he lived from 1579 to 1583. After obtaining a doctorate in theology from the university at Toulouse in 1581, he went on to Paris, then to Oxford in 1583, where he lectured on philosophy.

There was no Inquisition in England, which had broken away from the Roman Catholic Church during the reign of Henry VIII. But Bruno's contentiousness got him into trouble nevertheless. At that time, Oxford was a bastion of Aristotelian thought. Students were fined for disagreeing with Aristotle, and could be expelled if their views deviated from Aristotle's too sharply. Bruno nevertheless allowed himself to become involved in disputes during which he violently attacked the Aristotelian views of certain Oxford scholars. He was soon forced to leave the university.

He then went to London, to Paris, and finally to Germany, where he lectured at the universities of Marburg, Wittenburg, and Helmstedt. From there he went to Frankfurt, and then to Zurich. In 1591, a wealthy Venetian nobleman named Giovanni Mocenigo, an admirer of Bruno, offered the philosopher a position as a tutor. Bruno accepted, returning to Italy for the first time in many years.

It was an unfortunate decision. Bruno and Mocenigo did not get along, and the nobleman soon became alarmed by the heretical remarks that Bruno often made. In 1592, on the advice of his confessor, Mocenigo denounced Bruno to the Inquisition.

The trial never reached a conclusion. Before judgment was passed, Bruno was extradited to Rome. After being allowed to languish in the dungeons of the Roman Inquisition for six years, he was tried again. On February 16, 1600, he was burned at the stake, and it was decreed that his books be publicly burned and their titles placed upon the Index of books that Catholics were forbidden to read.

Sixteenth-century documents indicate that Bruno's inquisitors found some eight heretical propositions in his books. No document listing these propositions exists today. One can reasonably speculate, however, that one of the heretical ideas was Bruno's doctrine that human beings lived in an infinite universe containing an infinite number of worlds, some of which were populated by other human beings.

Bruno believed that the universe was infinite both in space and in time. Not only did it have no boundaries, it was eternal. Time stretched into the infinite past and into the infinite future. The sun was only one of an innumerable number of stars. Anything else was inconceivable. The mind could not help but think of the universe as infinite, and only an infinite universe could emanate from an infinite God.

According to Bruno, God had no choice but to create an infinite universe. Anything else would represent only a partial use of His infinite power. Bruno therefore rejected the Stoic conception, which postulated an infinite void beyond a finite world. In fact, in Bruno's universe, there was no void; an infinite material ether existed in the spaces between the worlds. This ether coexisted with God, who was omnipresent in infinite space. In *De l'infinito universo*, Bruno wrote:

> Thus is the excellence of God magnified and the greatness of his kingdom made manifest; he is glorified not in one, but in countless suns; not in a single earth, a single world, but in a thousand thousand, I say in an infinity of worlds.

It is not difficult to see that such views would be disturbing to the ecclesiastical authorities. According to Church doctrine, the world was created at a definite moment in time. It had not, as Bruno claimed, existed eternally. Furthermore, the idea of other inhabited worlds created serious problems. There had been but one Fall and one Redemption. It was from Adam and Eve that human beings had inherited original sin. If there were other earths with other inhabitants, how could they participate in the cosmic drama that had been enacted on earth? One had either to assume that the inhabitants of these worlds were all descended from Adams and Eves, and that Christ was crucified on all the worlds or, alternatively, that the people on the other earths possessed no original sin and had no need of redemption. But either assumption would be heretical.

To be sure, the idea of infinite worlds was not the only heretical doctrine that Bruno expounded. He believed in reincarnation, and taught that the earth—and indeed the universe itself—possessed a soul. Nevertheless, it is the concept of infinite worlds we think of when his name is mentioned today. Perhaps, as Bruno insisted, the idea of an infinite universe does have a certain appeal to the human mind. Contemporary scientists would not go so far as to maintain, as Bruno did, that it was impossible to conceive of a universe that was not infinite. Yet, Bruno's conception of infinity does capture our imaginations to a greater extent than his other, sometimes recondite, sometimes mystical, doctrines.

It certainly captured the imagination of the next century. There are numerous references to the doctrine of the plurality of worlds in the writings of John Donne, in Robert Burton's *Anatomy of Melancholy*, in John Milton's *Paradise Lost*, in the writings of the French essayist Michel de Montaigne, and in the fantastic science fiction novels of Cyrano de Bergerac.

By the end of the seventeenth century, the idea of the plurality of worlds had become commonplace. In 1734, the English poet Alexander Pope summed it all up in his *Essay on Man*:

> Through worlds unnumbered though the God be known,
> 'Tis ours to trace him only in our own.
> He, who through vast immensity can pierce,
> See worlds on worlds compose one universe.
> Observe how system into system runs,
> What other planets circle other suns,
> What varied Being peoples every star,
> May tell us why Heaven has made us as we are.

Belief in the plurality of worlds did not stem solely from the writings of Bruno. The astronomical discoveries that Galileo later made with a telescope also served to popularize the idea. However, if the Church's intention in burning Bruno in 1600 was to stop the spread of his ideas, the decision backfired. The execution only succeeded in publicizing the very doctrine it wanted to suppress.

THE SCIENTIFIC REVOLUTION

Although Bruno was an ardent Copernican, he was not a scientist, and he contributed little or nothing to the development of astronomy. It would be a mistake to associate him too strongly with the scientific revolution that was initiated by such individuals as Copernicus, Galileo, the Danish astronomer Tycho Brahe, the German astronomer Johannes Kepler, and the great Isaac Newton. In fact, Tycho, who was practically an exact contemporary of Bruno (he lived from 1546 to 1601, as compared with Bruno's 1548 to 1600), did much more to advance the Copernican cause than a dozen Brunos could have done. And he did it by slow, meticulous observation.

No truly accurate astronomical theory could have been built on the records of stellar observations that were available to astronomers in Copernicus's time. They contained too many errors. Some of the astronomical records contained data that had been collected by poor observers. Copying errors also took their toll; what had once been accurate data was sometimes passed on in a corrupted form. What was needed was a set of new, accurate observations upon which astronomers could rely. This is what Tycho provided.

Tycho, a Danish nobleman, lived an extravagant life. Granted a 2,000-acre island by the Danish king Frederick II, he built an observatory with royal patronage that contained not only the best (and most expensive) astronomical instruments, but also clocks, globes, sundials, allegorical figures and statues that turned on hidden mechanisms. Tycho also possessed game preserves and artificial fishponds, a printing press, a paper mill, and his own private prison. Like many members of the aristocracy of the day, he believed in living lavishly.

Tycho was by far the best astronomer of his era. He pushed naked-eye astronomy to its limits. He improved on almost every important astro-

nomical observation and calculated the length of the year to an accuracy
of less than a second. His data finally made calendar reform possible and
was the basis for the new Gregorian calendar, which was adopted in
1582. He observed the motions of the planets with an accuracy far
greater than any that had been achieved before.

Tycho was not a Copernican. He looked for some sign of stellar parallax, and when he could not observe this effect, he created a model with
Earth at the center of the solar system. In Tycho's system, the sun and the
moon revolved around Earth, while the five known planets revolved
around the sun. Tycho's system is mathematically equivalent to that of
Copernicus. In both, the planets revolve around the sun. And if one
wants to compute Earth's (or the sun's) orbit, it makes no difference
whether it is the sun or the earth that is assumed to be motionless; the
relative motion is the same in either case.

STEALING TYCHO'S OBSERVATIONS

After Tycho's death, Johannes Kepler, who worked as his assistant during the last year and a half of Tycho's life, was able to obtain the astronomical data that enabled him to work out the correct orbits of the
planets. While Tycho was alive, Kepler had been allowed to have only
limited information about his master's observations. Tycho had been willing to release the data only a tiny bit at a time. But once Tycho was dead,
Kepler helped himself. In a letter written just a few years later, he spoke
of "taking the observations under my care, or perhaps usurping them."

After years of arduous study and calculation, Kepler was able to deduce the laws of planetary motion upon which Newton was to build his
law of gravitation. Kepler found that theory could be made to agree with
Tycho's observations only if it was assumed that the planets moved in elliptical, not circular paths. Kepler's first law said that the planets moved
in elliptical orbits. His second law described how the velocity of a planet
changed at different points in its orbit. The third, which was finally announced in 1619, eighteen years after Tycho's death, related the distance
of a planet from the sun to the time that it took to complete a revolution.

Unlike Tycho, Kepler was a devoted advocate of the idea that the sun was the center of the solar system. It was his mathematical work that established the validity of the Copernican hypothesis once and for all. No experiments had yet been performed to demonstrate conclusively that the earth did move. That lay far in the future. However, Kepler had shown that it was possible to construct a coherent astronomical system that described planetary motion with near perfect accuracy. Furthermore, there was no need for epicycles and the other complicated mathematical devices that previously had been used by astronomers to patch up theories that never quite worked.

Like his predecessor Tycho, Kepler was much in demand as an astrologer. It was thought that his horoscopes were more accurate than those of his contemporaries, and they probably were. After he became court mathematician to the Holy Roman Emperor Rudolph II in 1601, one of his duties was the casting of horoscopes for Rudolph and other eminent persons. In this way the pseudoscience of astrology contributed to the progress of the scientific revolution by providing financial support.

During Kepler's time, ecclesiastical opposition to the heliocentric theory intensified. The Roman Catholic Church would not permit the printing of books that described the motion of the earth as something that was physically real until 1822. The Protestants sometimes objected to Copernican ideas even more than the Catholics did. However, Protestant religious authorities had no institution comparable to the Inquisition and were unable to exercise the kind of thought control that was common in Roman Catholic countries. So Kepler, who lived his life in a Protestant land, was able to carry out his work undisturbed. His only difficulties with religious authorities came about when he found himself having to defend his mother from charges of witchcraft.

When Katherine Kepler was accused of witchcraft in 1615, her son fought the charges and demanded that her persecutors send him copies of all the documents that were drawn up in connection with her case. Unfortunately, Katherine was foolish enough to offer one of the officials a bribe if he would quash one of the charges against her. This did not help her case. In December 1616, after deciding that flight was probably the best solution, Kepler and some other relatives took Katherine from the German principality she lived in to the Austrian city of Linz. But Katherine apparently didn't care for Linz; after living there for nine months, she returned home.

For some reason, nothing much happened at first. The court collected more evidence, but Katherine was left undisturbed until August 1620, when she was arrested and imprisoned. She remained imprisoned for fourteen months, and was finally released. It may be that her refusal to confess, even when she was threatened with torture, had something to do with this result. A large number of victims confessed once they were shown the instruments of torture. The sight could be quite terrifying. In those days, crushing the bones in the accused's thumb with a thumbscrew was often not considered to be torture at all, because it was so mild a punishment compared with other techniques that were employed. But Katherine wasn't fazed by the threats. She was willing to die, she told her accusers, and added that God would reveal the truth after her death. That might have made quite an impression, and of course that her son was the emperor's astrologer might also have been a factor associated with the eventual dropping of charges.

Kepler may have been the first person to write a science fiction novel about aliens. In *Somnium*, written in 1609 but not published until 1634, Kepler wrote of a voyage to the moon. The moon is divided into two hemispheres; the side always turned toward the earth is called Subvolva, while the far side is called Privolva. The two regions are inhabited by the relatively more civilized Subvolvans and the nomadic Privolvans, respectively.

The novel is no ordinary work of literature. It was written for other scientists in order to convince them of the validity of the Copernican system. For this purpose, Kepler inserts an astronomical treatise into the book. A young man named Duracotus has a mother who is a witch. For her son's benefit, she conjures up a wise spirit who gives a lecture on astronomy while also discussing such topics as the problems associated with space travel, the geography of the moon, and the moon's inhabitants.

When he was not writing science fiction, defending his mother against charges of witchcraft, or casting horoscopes, Kepler continued his scientific work. A book of tables of planetary motions based on Tycho's observations and his theory of elliptical orbits was published in 1627. The accuracy of these tables played a significant role in causing the idea of a sun-centered solar system to gain acceptance among scientists. At the same time, through the efforts of Galileo and others, it became known to a larger public as well. As early as 1611 John Donne, admitting that Copernican ideas "may very well be true," added that they were "creeping into every man's mind." During that same year, in his poem "An Anatomy of the World," he asserted that the "new Philosophy calls all in doubt."

He was referring, of course, to the new cosmology that had removed the earth from the center of the universe and made it into a planet.

GALILEO'S TELESCOPE

In 1609, Galileo heard that a "certain Dutchman"* had constructed an instrument that magnified the images of distant objects. Within a few months he had determined how such a device might be constructed and had made a telescope of his own. His first one was an eight-power telescope. This was quickly followed by one that produced a magnification of about twenty, and then by a thirty-power instrument. After he had made the last, Galileo began to use it to study celestial phenomena. He seems to have been the first person to have had the idea of turning a telescope upon the stars. Once he began, he immediately began to make a whole series of important discoveries. Galileo discovered four of the moons of Jupiter. He observed sunspots and found that the planet Venus had phases like those of the moon.

Galileo was an ardent supporter of the Copernican system and made much of these discoveries. But, although they made the heliocentric theory seem more plausible, they really didn't prove its validity. The fact that Jupiter had satellites showed that there were bodies in the solar system that did not revolve around Earth. However, this did not demonstrate that Earth revolved around the sun. Similarly, although the existence of the phases of Venus was inconsistent with the old, pre-Copernican sys-

---

* The German-Dutch lens grinder Hans Lippershey.

tems, their discovery did not contradict Tycho's theory, which also put Earth at the center of the solar system.

In fact, Galileo's astronomical ideas actually represented a step backward from those of Kepler. Galileo would not accept the reality of Kepler's ellipses. He maintained that since circular motion was the most natural, the planets had to have circular orbits. But Galileo did not concern himself very much with details. In his book *Dialogue Concerning the Two Chief World Systems*, published in 1632, he speaks of circular orbits without making any mention of the fact that the Copernican system required the use of epicycles and eccentrics under such circumstances.

Yet Galileo probably did more than Kepler to further the cause of the new astronomy. Kepler wrote primarily for other astronomers, and his ideas were slowly disseminated within the relatively small scientific community. News of Galileo's ideas and of his astronomical discoveries, on the other hand, spread quickly throughout Europe. This was obviously his intention; for example, he wrote the *Dialogue* in Italian rather than Latin in order to ensure that his book reached the widest possible audience.

In 1633, the year after the *Dialogue* was published, Galileo was summoned to Rome, where he was brought to trial and forced to recant. Nevertheless his influence was immense. The *Dialogue* was translated into Latin in 1835. Once that was done, it could be read by educated individuals in every European country. If Galileo's trial and recantation had any effect, it was to make it unfashionable *not* to be a Copernican.

The *Dialogue* did little to establish the truth of the Copernican system. Galileo was able to make the idea of a heliocentric solar system seem plausible, but he didn't advance the theory and none of his arguments constituted proof. Nevertheless, Galileo's book was influential and persuasive. His arguments effectively demolished the Aristotelian cosmology upon which the old astronomy had been based. Furthermore, he showed that the idea of a moving earth was perfectly reasonable from the standpoint of physics. He pointed out, for example, that the idea that an object dropped from a height onto a moving earth would not fall vertically was wrong. Anti-Copernicans had argued that such an object would lag behind an earth that was rapidly rotating under it, and would seem to move to the west. Galileo pointed out that it would participate in the earth's motion and move in a vertical path. He supported his argument by giving the analogy of an object dropped from the mast of a moving ship. Such an object would fall directly to the deck; it would not lag behind the ship. Similarly, he explained that the idea that a moving earth would create perpetually strong winds was false. The air was carried along with the earth as it rotated.

Some of the ideas advanced in the *Dialogue* are terribly wrong. Though Galileo knew that the atmosphere would be carried along with the earth, he seems to have believed that this would not always be true for the oceans. He tried to argue that the tides were the result of the earth's rotation. According to Galileo, the combined effects of the earth's rotation and its movement around the sun caused the seas to slosh back and forth in their basins causing sea levels to rise and fall. In fact, the tides are caused by the gravitational influence of the moon and, to a lesser extent, the sun. This was something that Kepler had guessed.

But perhaps Galileo can be forgiven for his mistaken theory of tides. In his day, little was known about gravity. Until Newton produced his theory of gravitation, no one would be able to conclusively demonstrate precisely how tides were produced. All in all, one has to conclude that, in spite of its defects, *Dialogue Concerning the Two Chief World Systems* is not only a great achievement but also one of the most effective propaganda weapons in the history of science. It was Galileo, not Kepler, who did the most to win acceptance for the heliocentric theory.

AN INFINITE UNIVERSE?

When Galileo turned his telescope upon the sky, he discovered numerous new stars too faint to be seen by the naked eye. When more powerful telescopes were constructed, astronomers discovered yet more stars. This soon gave rise to the idea that the number of stars must be limitless. If the stars all had about the same intrinsic brightness (an assumption that is *not* true, but which seemed reasonable at the time), then the new dim stars had to be farther away from the earth than those seen by a naked-eye observer. Since there was no reason to think that this process would ever come to an end, it was conceivable that the universe was infinite.

This line of reasoning is not perfectly compelling. It was also possible that all the stars were contained in a finite, though immensely large, sphere. Nevertheless, by the end of the seventeenth century, the idea of an infinite universe was widespread. Even Galileo, who had something of a horror of infinite magnitudes, was inclined toward this opinion.

In the *Dialogue*, Galileo states that it has not been proved whether the universe is infinite or finite. He then states that he is willing to accept a finite universe for the sake of argument. The matter is then quickly dropped. However, Galileo expressed himself somewhat differently in a letter he wrote in 1639 to Fortunio Liceti, a professor at the University of Padua. In this letter, Galileo stated again that he didn't believe that the matter could be conclusively settled. And then he added that he was inclined toward the idea of an infinite universe himself.

Galileo's reasoning about this matter is revealing. He wrote that he could not conceive either of a finite and bounded universe, or of one that was infinite and unbounded. Then he added that the fact that the infinite could not be comprehended by mankind's finite intellect caused him to prefer the latter possibility. Being baffled by the incomprehensible was to be preferred to finding oneself incapable of understanding the finite. The difference is striking between this point of view and the ideas that had been expressed by Bruno. To Bruno, the infinite had not seemed baffling at all.

THE LAW OF GRAVITY

In his great book *Philosophiae naturalis principia mathematica* (The Mathematical Principles of Natural Philosophy), which was written in Latin and published in 1687, Isaac Newton was able to show that celestial motions could be explained if one assumed that astronomical bodies attracted one another with forces that were inversely proportional to the square of their distance. For example, if the distance were to be increased by a factor of 2, the force between two bodies is ¼ as large, because 4 is the square of 2 (that is, $2 \times 2 = 4$). Similarly, if the distance were to be increased by a factor of 3, then the gravitation attraction would be ⅑ as great (because $3 \times 3 = 9$).

Newton's primary contribution wasn't the invention of an inverse square law of gravitation, however. His contemporaries, the English scientist Robert Hook and the architect Christopher Wren also guessed that the law of gravity would have to have this form. The reason that Newton is generally given credit for discovering the law of gravity is the fact that he was the only one who possessed the mathematical skills necessary to show that if gravity did decrease as the square of the distance, then it followed that the planets would have orbits corresponding to Kepler's ellipses.

Newton was never in a hurry to publish his results. According to an often-told story, one day in 1684 he received a visit from the astronomer Edmond Halley. Halley asked Newton what kind of orbit a planet would have if the force of gravity did vary according to an inverse square law. Newton immediately replied that it would be an ellipse. Asked by the amazed Halley how he knew, Newton replied, "I have calculated it." But when Halley asked him for the calculation, Newton was unable to find it among his papers.

As Newton's biographer Richard Westfall points out, Newton's inability to find the calculation was probably a charade. He probably wanted to check his calculations—which had been done many years previously— before he allowed anyone else to see them. But in any case, Halley soon received from Newton a nine-page treatise entitled *De motu corporum in gyrum* (On the Motion of Bodies in Orbit). Halley immediately recognized its importance.

Once he had composed *De motu*, Newton was seized with the desire to write a book dealing with the problems of motion and the dynamics of the solar system in a more comprehensive way. He completed his monumental *Principia* in eighteen months and sent it to the Royal Society in 1686. At this point, Halley again played a leading role. It was he who proposed the resolution that the society should publish the book. This action almost cost Halley his job. At the time he was clerk to the society, and employees of his stature were not supposed to propose major projects. In the end, all turned out well. The *Principia* was published in 1687, and Halley remained employed.

NEWTON'S ARGUMENT
FOR AN INFINITE UNIVERSE

Newton's major concern in the *Principia* was in giving accurate mathematical demonstrations, not in speculating about such matters as the infinity or the finiteness of the universe. Nevertheless, it is clear he believed the universe to be infinite. In his book *Opticks*, which was written in English and intended for a wider audience, Newton described the universe in a way that was reminiscent of Bruno's descriptions of it. Like Bruno, Newton believed that infinite space was filled with an omnipresent God. In one edition, he went so far as to describe the infinite universe as "God's sensorium," a statement that was ridiculed by the German philosopher Gottfried Wilhelm von Leibniz. Newton must have winced when he heard about Leibniz's laughter. At the time, the two men were involved in an acrimonious dispute over which of them had first invented the calculus.

In *Opticks*, Newton produced no arguments for an infinite universe. For this, we must turn not to his published work but to a letter. Writing to the English clergyman Richard Bentley in 1692, Newton argued for an infinite universe. If the universe were finite, he said, gravity would cause all the matter in the universe to collect at its center. In an infinite universe, on the other hand, any individual body would experience gravitational forces in all directions. If matter was more or less evenly distributed, these forces would be more or less equal, and it would not be pulled in any particular direction.

It seems simple enough that if the universe were finite, a body near the edge of the universe would experience little or no outward force, while it would experience gravitational forces pulling it inward. Nevertheless, the argument is incorrect. It appears that even great minds can become confused when they consider infinite numbers of objects. If there were an infinite number of stars spread out more or less evenly through space, all that would be needed to cause them to clump together would be a tiny excess of matter in one region or another.

Modern astronomical observations have revealed that our universe has just such a clumpy structure. Stars are clustered together in galaxies.

Galaxies have come together in clusters of galaxies. Galaxies, however, remain stable over periods of many billions of years. Further collapse is prevented by the galaxies' rotation. Individual stars revolve around the galactic center in the same way that the planets in the solar system revolve around the sun.

In Newton's day, of course, the existence of galaxies had not yet been discovered. Astronomers could observe only the stars in our own Milky Way galaxy. As far as they could tell, stars did seem to be more or less uniformly distributed through space. Thus Newton's argument must have had a certain plausibility that it does not possess today.

OLBERS' PARADOX

By the end of the seventeenth century, no one had found any compelling arguments in favor of an infinite universe. Many individuals had concluded this was something that couldn't be done. Conceiving of infinity was one thing; observing it was something else. After all, no one could look an infinite distance out into space or count an infinite number of stars.

But perhaps the universe wasn't infinite. In that case, could anyone find a convincing argument that would show it was finite? Halley thought that he had one. In 1720, he argued that if the universe really were infinite, then the sky would be as bright as the surface of the sun. In an infinite universe with an infinite number of stars, there would be no place in the sky you could look without having your line of vision fall on one star or another. An infinite number of stars would not produce an infinitely bright sky; the farther away they were, the dimmer they would appear to be. But the light they produced would still be enormous.

This argument did not originate with Halley. It had previously been stated by Thomas Digges and by the Swiss astronomer Jean-Philippe de Cheseaux. Ironically, neither Cheseaux nor Halley wound up getting credit for the idea. Today the argument is known as Olbers' paradox, after the German astronomer Heinrich Olbers, who restated it in 1826.

Olbers' paradox is not easy to dispose of, though at first glance it appears that it might be. Today, scientists believe that the universe began in a big bang that took place approximately 15 billion years ago. If this is true, even if the universe is infinite, light from an infinite number of stars cannot reach Earth. At best, we can see only those that are less than 15 billion light years away. This follows from the definition of a light year: the distance that a ray of light travels in a year. The first person to suggest this solution to the paradox may have been Edgar Allan Poe. Poe suggested in 1848 that when we look out into space and back in time, we might be seeing the darkness that existed before the stars were created.

But if the universe began with a big bang, a new problem arises. Wherever we look in the sky, we should see the light from the big bang fireball. After all, the fireball was something that existed everywhere. It filled the entire universe. But this is a problem I will have to return to later, when I discuss the expanding universe.

4

THE INFINITELY SMALL

THE UNIVERSE (which others call the Library) is composed of an indefinite and perhaps infinite number of hexagonal galleries. . . ." So begins Borges's short story "The Library of Babel." In each hexagon of this infinite library (the narrator soon tells us that he prefers to believe that it is infinite), there are twenty shelves, each containing thirty-five books of four hundred and ten pages. The library, which has existed for all eternity, and which will continue to exist throughout the infinite future, contains every possible four-hundred-and-ten-page book. Every conceivable combination of letters exists in some book. As a result, most of the volumes are unintelligible. For example, there is one book in which the letters MCV are repeated from the beginning to the end. Another is made up of what appears to be a random assortment of letters, except that the words "Oh time thy pyramids" are found on the next to the last page.

Some of the books are less senseless than others. Borges's narrator tells us of a book that had been found five hundred years before his time that contained nearly two pages of meaningful lines. At first it was thought that these lines were written in Portuguese, or at least that was what a "wandering decoder" pronounced them to be. Others were not so sure; they maintained that the words were Yiddish. Eventually scholars established the true nature of the language. It was a Samoyedic Lithuanian dialect of Guarani, with classical Arabian inflections.

In the second paragraph of the story, the narrator parodies the medieval dictum about God, asserting that "The Library is a sphere whose exact center is any one of its hexagons and whose circumference is inaccessible." At the conclusion, he again argues that the universe (the Library) is infinite, and attempts to meet the objections of those who point out that the number of possible four-hundred-and-ten-page books, though very large, is not infinite. "The Library is unlimited and cyclical," he maintains. If it were possible to travel through it in any direction, he says, one would eventually come to a place where the same volumes began to be repeated in the same order.

Borges's conception of the Library recalls the Stoic theory of cosmic cycles, with the exception that it is now space, rather than time, which is cyclical. In addition, Borges seems to have anticipated modern cosmologists who point out that it is possible, though not very likely, that our universe could have such a structure. However, I find an idea that Borges expresses in a footnote at the very end of the story to be even more interesting. This infinite Library, he says, could be replaced by a single book containing an infinite number of infinitely thin pages.

This idea is a variation of a suggestion that was made in the early seventeenth century, when the Italian mathematician Bonaventura Cavalieri had conceived of solid bodies as being made up of infinite numbers of planes. According to Cavalieri, a cylinder could be thought of as being made up of an infinite number of infinitely thin circular disks, while a cube was composed of an infinite number of infinitely thin squares. Of course, the ancient Greek mathematicians had conceived of a line as being made up of an infinite number of points, so the idea wasn't entirely new. Cavalieri was original, however, in that he found ways to use the idea for making mathematical calculations.

But such considerations elicit a whole series of questions. For example, is it really possible to speak of infinitely small quantities? Could one really add them up in such a way to get a finite result? Exactly what would such things be? Would infinitely small quantities be equal to zero in some sense? Or would they have dimensions that were larger than zero, but smaller than any finite quantity? When one divides a number by infinity, is the result zero? Or is it some infinitesimal quantity?

FALLING BODIES

According to a story first recounted by Galileo's disciple Vincenzo Viviani long after Galileo was dead, Galileo once gave a demonstration in
which he dropped two weights from the Leaning Tower of Pisa. He supposedly did this to demonstrate that both the heavier and the lighter
weight would strike the ground at the same time.

Today, scholars agree that it is unlikely that any such demonstration
ever took place. If such an experiment were performed, the weights
would not strike the ground at exactly the same instant. For if a light and
a heavy weight are dropped from a height, the heavy one will reach the
ground first. The difference in the times that it takes the two objects to
fall will not be great. However, the heavy one will be impeded less by air
resistance. Only in a vacuum do all objects fall at the same rate.

The behavior of falling bodies was a subject that greatly interested
Galileo. However, he did not experiment with objects by observing their
vertical fall. If an object is dropped, everything happens too quickly. You
cannot tell whether the object instantly acquires some velocity the moment it is released, or whether it begins with zero velocity and then begins to move faster and faster. If the object is accelerating during the fall,
how can we determine how fast it is falling at any given moment?

Before Galileo could perform experiments on the behavior of falling
bodies, he had to find some way to "dilute" the effects of gravity. He accomplished this by making use of inclined planes. If a ball is allowed to
roll down a plane that is tilted to some degree above the horizontal, it will
still be subject to the effects of gravity, but its "fall" will take place in
slow motion. For example, if a ball is dropped from a height of six feet, it
will reach the ground in a little over six-tenths of a second. But it will
take almost nine seconds to roll down a plane that is inclined at a four-
degree angle.

Galileo was one of the first scientists to perform experiments in order
to study the actual behavior of objects in the natural world. In his day,
when Aristotelian thought dominated the universities, such a thing was
quite unusual. If a "scientist" wanted to know, for example, how a falling

body would behave, he would attempt to find out by looking to see what Aristotle said about the matter, or use Aristotelian principles to make deductions. Thus it was generally thought that heavy bodies always fell faster than light ones: Aristotle said they did.

Though Galileo was a great scientist, he didn't approach experimentation with the rigor of a modern physicist. In fact, he was not above engaging in a little pious fraud. He sometimes reported results that were too good to be true; the reported accuracy was greater than anything he could have achieved with the instruments available at the time. On other occasions, he discussed experiments that scholars think he did not really perform. Apparently he believed that if he knew how an experiment would turn out, there was no need actually to do it.

He did manage to reach correct conclusions, however. After struggling with the problem of falling bodies for years, he finally succeeded in finding the mathematical formulas that described an object's fall. He found that the velocity at any given moment was proportional to the time it had been falling, and that the distance an object fell was proportional to the square of the time. For example, an object that had been falling for three seconds would be moving three times as fast as one that had been dropping for only one second. And in three seconds, an object would drop nine times as far as it would in one second (because $3^2 = 9$).

To most modern readers, discovering the laws that govern the behavior of falling bodies probably doesn't sound like a difficult problem. But in fact it was. In the first place, it was generally assumed in Galileo's day that, at any moment, the velocity of a falling object was proportional to the distance it had fallen. Galileo himself subscribed to this error for many years, believing, for example, that a body that had fallen five feet would be moving five times as fast as one that had fallen one foot. After he realized that this assumption was wrong, he had to find the correct laws of accelerated motion. A falling body is accelerated because it does not move at a constant speed. At any instant of time, it moves a little faster than it had been moving at a slightly earlier instant. This kind of motion is more difficult to analyze than a case where the velocity always remains the same.

INSTANTANEOUS VELOCITY

When a scientific puzzle is solved, new, deeper questions often arise. After Galileo succeeded in finding a correct description of the behavior of falling bodies, he found himself confronting a new problem. It is easy to understand and measure average velocities. If an object falls sixteen feet in one second, its average speed was sixteen feet per second. However, the object will also have an instantaneous velocity at any given time. After a tenth of a second, it will be moving at some specific speed. After four-tenths of a second, it will be moving at some other velocity. But it was anything but obvious to Galileo how instantaneous velocity should be defined.

In order to see why this is a problem, consider the following thought experiment. Suppose we drop a baseball from the Leaning Tower of Pisa. We want to determine exactly how fast it is falling after exactly one second has elapsed. So we photograph it at one instant, and then again one-tenth of a second later. The photographs show how far the ball has fallen during that period of time. We can then divide the distance traveled in feet by the elapsed time in seconds to obtain a speed (feet per second). Unfortunately, this only gives the average velocity during that one-tenth of a second. So we repeat the experiment. This time we photograph the ball at two instants that are a thousandth of a second apart. But again we obtain only an average velocity. After all, the ball was moving faster after that thousandth of a second had elapsed than it was at the beginning of that period of time. Even if we photograph the ball at two instants a billionth of a second apart (this could not be done in reality, but in thought experiments all things are possible) we still only get an average velocity. Obviously, this would be an accurate result, but it would still not be precisely equal to the ball's instantaneous velocity at a time of exactly one second.

The problem baffled Galileo. A falling object must have a definite speed at any point in its fall. Average velocities can be defined for shorter and shorter periods of time. But if we want to know what is happening at some particular instant, the method fails. A ball will fall zero feet in zero seconds. This doesn't give us a velocity at all. The number %, after all, is meaningless. We have a situation that is reminiscent of Zeno's paradox "The Arrow," in which Zeno argued that motion was impossible because at any given instant an arrow was motionless.

INFINITESIMALS

Perhaps the way to deal with the concept of instantaneous velocity was by saying that the ball falls some infinitely small distance in an infinitely small period of time. As we shall see, this is exactly what was done. There were doubts as to whether this really solved the problem, since it was not so easy to say exactly what an infinitely small quantity was. However, the idea led to the development of powerful new mathematical methods that could be used to describe the behavior of bodies that had constantly changing velocities.

These methods were not available immediately. Before that could happen, it was necessary that the ground be prepared. This happened very slowly. It took scientists many years to develop workable methods. The concept of the infinitely small was so new and strange that scientists had to learn to be comfortable with it.

One of the first to make use of infinitely small quantities, or infinitesimals, as they later came to be called, was Cavalieri, a Jesuit mathematician who considered himself a disciple of Galileo. Cavalieri showed that it was possible to calculate the volumes of solid bodies by making the assumption that they were made up of very great numbers of infinitely thin sheets that he called "indivisibles." I have already given two examples of this procedure, noting that a cylinder could be viewed as composed of an infinite number of infinitely thin disks.

Cavalieri could never quite say exactly what an indivisible was, and his methods were much criticized by his contemporaries. He could not explain how a body of finite size could be formed from an infinite number of elements, and he sometimes described his innovation as nothing more than a pragmatic procedure that allowed one to avoid using more complicated mathematical methods. But even though he could give no clear explanation of what it was that he was doing, he made one of the first steps toward the concept of the infinitesimal.

Cavalieri's methods captured the interest of a few thinkers, including the French mathematician and physicist (and author of the famous *Pensées*) Blaise Pascal. According to Pascal, the method of indivisibles was perfectly consistent with classical Greek geometry; what could be demonstrated with one method could also be demonstrated with the other. Any mathematician who wanted to call himself a geometer, Pascal said, was obliged to accept Cavalieri's technique.

Pascal was not the only mathematician to make use of Cavalieri's indivisibles. Others included the French mathematician Pierre de Fermat and the Oxford mathematician John Wallis. Fermat was a counselor for the Toulouse parliament who pursued mathematics for his own amusement. He was in the habit of writing down his mathematical discoveries in the margins of books. But his work on infinitesimals remained unknown, as he didn't bother to publish it. The work of Wallis, who was one of the founders of the Royal Society, was much more significant. He did publish, and his work was intensely studied by Newton.

THE LAST OF THE MAGICIANS

Newton was not the first of the age of reason," said the British economist John Maynard Keynes in an essay written for the Newton tercentenary celebrations in 1942. "He was the last of the magicians, the last of the Babylonians and Sumerians, the last great mind which looked out on the visible and intellectual world with the same eyes as those who began to build our intellectual heritage rather less than 10,000 years ago." Keynes went on to assert that Newton looked at the natural world as a riddle created by God, one that could be solved by finding the "mystic clues which God had laid about."

Keynes certainly had a point. Although Newton is remembered today as a great scientist, during his lifetime he spent more time conducting alchemical experiments than he ever spent on his work in physics or his mathematical discoveries. Upon his death, Newton left manuscripts containing over a million words of writing on alchemy and theology. Only a small portion of this writing has ever been published.

Where Galileo was charged with heresy for championing a scientific theory, Newton really was a heretic. He denied the Christian Trinity, and though he believed that Christ had been more than a man, Newton believed him to be subordinate to God the Father. Naturally, he kept these beliefs secret. If they became known, he would not be prosecuted, but he would lose his fellowship at Cambridge and be ostracized from British society. It is significant that when the British Parliament passed a bill in 1673 allowing public worship by Protestant dissenters, this privilege was specifically denied to "any person that shall deny in his Preaching or Writeing [sic] the Doctrine of the Blessed Trinity."

It is doubtful that Newton's alchemical and theological preoccupations hindered his scientific work, however, for he pursued science with an uncommon intensity. When he was working on a scientific or mathematical problem, he would often forget to eat and sometimes forget to sleep as well. On more than one occasion he spent the entire night on a problem, unaware of the passing time. He was also oblivious to the fact that some of the things he did might cause bodily injury. When he was working on problems concerning vision and the nature of light, he looked into the sun in order to observe the afterimages that appeared when he looked away. On another occasion, he inserted a bodkin—a needlelike instrument—between his eye and his eye socket. He wanted to see what would happen when he put pressure on the eyeball in order to alter the curvature of the retina.

FLUXIONS

Newton was the inventor of the mathematical technique that is now called the differential calculus. Characteristically, Newton neglected to publish his work. As a result, he was to become involved in a bitter priority dispute with Leibniz, who discovered the differential calculus independently. The battle raged on for years, and continued even after Leibniz's death in 1716. Both Newton and Leibniz were accused of plagiarism. Members of the British camp charged that Leibniz had stolen Newton's idea. Continental mathematicians, who generally supported Leibniz, charged that Newton had developed his method of "fluxions" in imitation of the work of his German rival.

Today, credit is given to both men. However, it is Leibniz's version of the calculus that is most often used. Though the methods of Newton and Leibniz were equivalent, Leibniz succeeded in putting the calculus in a form that made calculations easier to carry out, and his superior notation is still used today. Even Newton's terminology has died out. Today, no one refers to the calculus as the method of fluxions, except in works that deal with the history of mathematics.

Like many mathematical and scientific ideas, the differential calculus is based on a very simple one. Newton solved the problem of instantaneous velocity by defining it as the ratio of two vanishingly small quantities, two infinitesimals. He then generalized the method by noting that it was possible to talk of the ratio of any two infinitesimal quantities. The idea was new because, although many mathematicians had done calculations in which infinitesimals were "added up" to produce a finite sum, no one had thought to see what would happen if one were divided by another.

If all Newton had done had been to create a definition, little or nothing would have been accomplished. After all, though they had proven to be useful, infinitesimals were still suspect quantities. No one had, as yet, found a way to adequately define them. What Newton did was to find ways to use infinitesimals to perform calculations. It was the calculus, for example, that enabled him to determine that if gravity followed an inverse square law, then the orbit of the planets would be ellipses. And it was the calculus that allowed him to obtain the important result that the gravitational force created by a massive body was the same as the force that would be created if all of its mass were concentrated in its center. Without this result, Newton's law of gravitation would have been cumbersome indeed.

The calculus represented a great advance because it gave scientists a method for dealing with the behavior of bodies, such as falling weights or orbiting planets, that did not move at constant velocities. Furthermore, it could be used to describe the behavior of *any* quantity that varied in time. For example, Newton used it to determine the manner in which hot bodies cooled. A hot object does not lose heat to its environment at a constant rate. A red-hot piece of iron gives off much more energy in any given period of time than one that has cooled to the point where it is only warm. The calculus must be used to describe what is happening.

Furthermore, the calculus can be used to describe quantities that vary in space as well. A magnetic field, for example, is strongest at points near one of the poles of a magnet, and weaker at distances farther away. Until Newton and Leibniz made their discoveries, there was no simple way to deal with such varying quantities mathematically. The invention of the differential calculus must be considered to be one of the greatest mathematical discoveries of all time.

BUT WHAT IS AN INFINITESIMAL?

Instantaneous velocity could be defined as the ratio of two infinitesimals, a vanishingly small distance divided by a vanishingly small time. That had proven to be an invaluable idea. But what, exactly, were these quantities? No one could say, including Leibniz and Newton.

Newton made several different attempts to define infinitesimals, but none were very accurate. On one occasion he referred to them as ultimate indivisibles, on another as "evanescent divisible quantities," and on yet another as "nascent increments." When writing the *Principia*, he derived many of his results by using the calculus, and then substituted geometrical demonstrations whenever possible. He obviously wanted to present his work in a form that would seem more rigorous than it would have appeared if he had given the calculations that he actually used.

Leibniz did no better. He didn't waver back and forth when making definitions the way Newton did. He generally spoke of quantities that were "vanishingly small" or "infinitely small." But when after twelve years of work he published his first paper on the calculus, he found himself unable to give a clear explanation. Leibniz's friends, the Swiss mathematicians Johann and Jakob Bernoulli, described Leibniz's account of the calculus as "an enigma rather than an explication." Five years later Leibniz wrote another paper in which he described infinitesimals as "fictitious numbers." But this hardly seemed to improve matters. Leibniz eventually gave up on attempts at definition. In 1695, he wrote that "overprecise" critics should not deter anyone from using what had proven to be a very useful method.

Some of Leibniz's explanations sound baffling indeed. In one of his letters, he said that the infinitely small was not a "simple and absolute zero, but a relative zero." It was an evanescent quantity, which retained the character of that which was disappearing. On some occasions, he defended the use of a "form without magnitude." On others, he denied that there were really infinite or infinitesimal magnitudes.

Some mathematicians, including Wallis and Johann Bernoulli, tried to define the infinitesimal as the number 1 divided by infinity. But this didn't work, either. Infinity is not a definite number in the way that numbers like 1 or 5 or 783,692 are. Today, mathematicians generally consider the quantity $1/\infty$ (∞ is the mathematical symbol for infinity) to be undefined. Bernoulli went on to write the first text on the calculus. In this book, one finds such cryptic remarks as, "A quantity which is diminished or increased by an infinitely small quantity is neither increased nor decreased." Bernoulli also gave analogies, but these were hardly more illuminating. At one point he says that infinitely large quantities were like astronomical distances while the infinitely small ones were like the tiny creatures that were then first being seen with microscopes. This isn't even a good analogy. Astronomical distances can be large, but they are hardly infinite, and the single-celled creatures that can be seen by a microscope are certainly not infinitely small.

It appeared that the calculus was based on rather shaky foundations. To make matters worse, certain problems in physics sometimes made it necessary to use second-order infinitesimals. Quantities of the second order are encountered when relating acceleration to distance traveled, for example. As we have seen, infinitesimals were used to define instantaneous velocity. Acceleration can be defined as the rate at which velocity changes. Acceleration therefore involves infinitesimals of infinitesimals. But what could second-order infinitesimals be? They looked like quantities that were infinitely smaller than quantities that were already infinitely small. Nor did it end there. It was possible to write mathematical expressions that contained infinitesimals of even higher orders. What were those?

AN INFIDEL MATHEMATICIAN

In 1734, the British philosopher Bishop George Berkeley published a book entitled: *The Analyst Or a Discourse Addressed to an Infidel Mathematician.* Wherein It Is Examined Whether the Object, Principles, and Inferences of the Modern Analysis Are More Distinctly Conceived, or More Evidently Deduced, than Religious Mysteries and Points of Faith. "First Cast the Beam Out of Thine Own Eye; and Then Thou Shalt See Clearly to Cast the Mote Out of Thy Brother's Eye."* In this book, he charged that the calculus was based on illogical foundations. Speaking of infinitesimals, he said, "They are neither finite quantities, nor quantities infinitely small, nor yet nothing. May we not call them the ghosts of departed quantities . . . ?" When one examined expositions of the calculus, Berkeley said, it was to discover "much emptiness, darkness and confusion; nay, if I mistake not, direct impossibilities and contradictions."

You might wonder why a philosopher who denied the existence of the material world should have been so concerned about the existence of mathematical quantities like the infinitesimal. The answer is that Berkeley was disturbed about the growing influence of the philosophies of mechanism and determinism in science. Fearing that this was a threat to religion, he counterattacked by focusing on what he believed to be the weak point of eighteenth-century science.

Numerous mathematicians attempted to answer Berkeley's charges. None were very successful. The nature of infinitesimals was indeed a mystery, and in some cases the answers only seemed to make matters worse. For example, the Swiss mathematician Leonhard Euler concluded that the ratio of two infinitesimals was zero divided by zero. The number

---

* The infidel mathematician was Halley.

%, he claimed, could be equal to anything. Euler was a great mathematician, but today this conclusion looks like nonsense. It appears that great minds could become befuddled when dealing with the infinitely small as well as with the infinitely large. This was not the only time that contemplation of the infinite led Euler into absurdity, incidentally. On another occasion, he concluded that the number −1 was greater than infinity. Infinity, he said, resembled zero in that it divided the positive from the negative numbers.

In 1784, the Berlin Academy of Sciences offered a prize for the best solution to the problem of the infinite. The announcement of the competition made specific reference to the fact that mathematics employed both the infinitely great and the infinitely small, and stated, "The Academy, therefore, desires an explanation of how it is that so many correct theorems have been deduced from a contradictory supposition."

Some twenty-three papers, most of which dealt with the calculus and with infinitesimals, were submitted. After examining the papers, the Academy concluded that none of them were completely satisfactory, and that they generally lacked clarity, simplicity, and rigor. Commenting, "The feeling of the Academy, therefore, is that its query has not met with a full response," the Academy nevertheless awarded a prize to the mathematician its committee thought came closest to fulfilling its intentions. This was the Swiss mathematician Simon L'Huiller, who attached a motto, "The Infinite is the abyss in which our thoughts are engulfed," to his paper "Elementary Exposition of the Higher Calculus."

Throughout the eighteenth century mathematicians struggled to find some way in which the calculus could be given a logical foundation. At best, they only succeeded in making matters more confusing. This fact was not lost on Voltaire, who remarked that the calculus seemed to be "the art of numbering and measuring exactly a Thing whose Existence cannot be conceived." Meanwhile, the French mathematician Jean Le Rond d'Alembert assured his students that it was indeed worthwhile to study these doubtful methods. The calculus, he told them, might seem illogical at first, but faith would eventually come to them.

Yet while confusion reigned, the calculus went from success to success. The work of Newton and Leibniz was extended. The calculus stimulated the creation of new branches of mathematics, and numerous problems in physics and mathematics were successfully solved. It was apparent that whatever the logical status of infinitesimals, the calculus gave correct results.

THE THEORY OF LIMITS

The calculus was finally put on a firm foundation in the nineteenth century. In 1821, the French mathematician Augustin-Louis Cauchy published *Cours d'analyse* in which he outlined a way to eliminate the troublesome concept of the infinitesimal. Cauchy accomplished this by demonstrating that the calculus could be based on the idea of *limit* instead.

We have already encountered this conception in the discussion of the paradoxes of Zeno. Consider the paradox of "The Dichotomy." As we saw in a previous chapter, Zeno claimed that a runner could never actually reach his goal, though he got closer and closer to it. If this were really true, it would nevertheless be possible to state the exact location of the point that the runner was trying to reach. Similarly, the sequence 1, ½, ¼, ⅛, 1/16, 1/32 . . . gets closer and closer to zero, but never actually reaches it. As the sequence is continued, the numbers become arbitrarily small, but there are no infinitely small quantities.

Newton's fluxions—which today are called derivatives—were originally conceived to be ratios of infinitely small quantities. Cauchy defined them in a somewhat different way. He took the ratio of two finite quantities, and then allowed each of them to become smaller and smaller. As they did, the ratio would approach some limit. A good example of this is Galileo's instantaneous velocity. Let us go back to the thought experiment of a falling body that I gave at the beginning of this chapter. We wanted to know how fast an object was falling exactly one second after it was dropped, and imagined that it was photographed at two instants, a tenth of a second apart. This only gave an average velocity during that tenth of a second. So we then imagined photographing the ball at instants a thousandth of a second apart, and then a billionth of a second apart.

None of the average velocities thus computed will be exactly equal to the instantaneous velocity, but they will get closer and closer to that number as the time interval is decreased. We might, for example, get average velocities equal to 4.7, 4.9, 4.99, 4.999 feet per second, and so on. They approach an instantaneous of 5 feet per second as a limit.

Cauchy thought that he had given the calculus a sound logical foundation. Indeed, he had the correct idea, but he was dealing with a difficult subject and made a number of mistakes when he attempted to apply the idea of limit in specific cases. His book contained a number of erroneous assumptions and invalid proofs. What was needed now was some way to make Cauchy's work rigorous. This was accomplished about fifty years later when the German mathematician Karl Weierstrass refined Cauchy's methods, showing at last that the calculus could be given a logical foundation. An explanation of how he accomplished this would be a little too technical for a book of this sort. I must therefore limit myself to saying simply that it was a great achievement.

Don't imagine that the concept of the infinitesimal had been done away with. As a matter of fact, physicists never stopped using infinitesimals and commonly write mathematical formulas in which one infinitesimal is equated to certain others. It provides a very intuitive way of looking at things, and physicists tend not to worry very much about mathematical rigor, as long as the methods used give the correct results.

In universities, it is common to find calculus being handled one way in the mathematics department and somewhat differently in the physics department next door. Mathematicians carefully define everything in terms of limits, while this is a concept that physicists often ignore. I first became aware of the difference as an undergraduate student when I was simultaneously taking my first college physics course and my first course in calculus. Some of my fellow students were having trouble learning calculus. So our physics professor recommended a book on the subject, *Calculus Made Easy* by Sylvanus Thompson.* I recall that this book defined a derivative as a quantity that was equal to a "little bit" of something di-

---

* As I write this, Thompson's book is still in print, even though some thirty-seven years have passed since I took undergraduate calculus.

vided by a "little bit" of something else. This is essentially the definition that was made by Newton.

THE INFINITESIMAL RETURNS

We can't criticize the physicists too harshly for their lack of rigor. During the twentieth century, the concept of the infinitely small was reintroduced into mathematics. A number of mathematicians began to suspect that it might be possible to construct a consistent theory of infinitesimals, and in 1934 the German mathematician Felix Klein suggested that this could be done if one abandoned the axiom of Archimedes, one of the axioms on which the number system is ordinarily based.

The axiom of Archimedes says that if we take any two numbers—let us call them A and B, and assume that A is less than B—and if A is then added to itself a sufficient number of times, the result will be a sum that is greater than B. For example, if A is 1 and B is 25, we will get a number that is greater than B if we add 1 to itself twenty-six times.

To non-mathematicians, this idea may seem too simple and obvious to bother about. But it is necessary to use it to treat the number system in a truly rigorous way. It should be apparent that the Archimedean axiom cannot be satisfied if infinitesimals are admitted into the number system. For no matter how many times an infinitesimal is added to itself, the sum will never be greater than 25, or even 1, for that matter.

You might think that getting rid of a standard axiom is not a legitimate procedure. This is definitely not the case. In Newton's time, no one would have thought of tinkering with standard axioms, but today matters are viewed quite differently. You are free to use any set of axioms you choose, as long as a consistent mathematical system results. Mathematics is not viewed the way it was three centuries ago.

During the 1960s, largely through the efforts of the mathematician Abraham Robinson, a new mathematical theory of infinitesimals was developed. In this system, called non-standard analysis, an infinitesimal is defined as a number that is greater than zero but less than any positive number. The system also makes use of hyperreal numbers, which are the sums of ordinary numbers and infinitesimals. For example, if one adds an infinitesimal to the number 6.32, a hyperreal number is the result.

A calculus based on hyperreal numbers isn't quite the same as the calculus in the days before Cauchy, but there are more similarities than differences. Some standard theorems have proofs that look a little different. However, the same results are obtained when calculations are performed. Neither the calculations in physics that make use of the calculus nor the mathematical systems that are based on the calculus would have to be changed one whit. Of course, this raises the question of whether or not non-standard analysis is going to be useful to mathematicians. Though it's an interesting theory, it appears that it has not yet produced any important new mathematical results.

Non-standard analysis does introduce new infinite numbers, which are not the transfinite numbers of Cantor. These new infinities are obtained when one divides an ordinary number by an infinitesimal. Thus it has enriched mathematicians' conception of the infinite to some extent. One suspects that Cantor would not approve if he were able to see what was being done. Cantor, for all his preoccupation with the infinite, would have nothing to do with the infinitely small. When another mathematician proposed using Cantor's transfinite numbers to work out a system that included infinitely small quantities, Cantor accused him of attempting to infect mathematics with "the Cholera-Bacillus of infinitesimals."

5

ATOMIC CATASTROPHE

I THINK I WAS IN JUNIOR HIGH SCHOOL when I first heard the question "What happens when an irresistible force meets an immovable object?" Even then, the question didn't particularly trouble me, even though it wasn't until many years later that I became aware of the answer: "It passes right on through."

Of course, the question is as facetious as the answer. There are no such things as infinitely strong forces, or objects that offer infinite resistance to movement. No one has ever observed any infinitely large quantity. When infinities are predicted by theories in physics, it is generally an indication that something has gone wrong, that the theory is based on invalid assumptions, or that it has reached its limits of applicability.

It is true that scientists sometimes speak of the possibility of an infinite universe. Even in this case, however, the infinite could not be directly observed. If the universe is approximately 15 billion years old, then it is impossible to observe any object that is more than 15 billion light years away. Light emitted by any galaxy or other object at a greater distance has not had enough time to reach us. An object at a distance of more than 15 billion light years could not affect us in any way. Gravity is the only significant long-range force in the universe, and it also propagates at the speed of light.

The observable universe, in other words, is finite. It is a sphere with a radius of 15 billion light years. To be sure, this sphere is perpetually expanding. When the universe is 20 billion years old, intelligent beings—if any exist—will be able to observe objects at distances up to 20 billion light years away. However, 20 billion light years is nothing compared with an infinite distance. When one speaks of the possibility of an infinite universe, it is not necessary to deal with quantities that are truly infinite. This is fortunate, because the equations of physics can only deal with quantities that are finite.

THE DISCOVERY OF THE ELECTRON

As it turned out, an encounter with the infinite contributed to the development of modern physics. By around 1890, the physicists thought they had everything figured out. The two major subdivisions of physics—mechanics and electromagnetism—seemed to explain all natural phenomena. Mechanics was the field of physics that dealt with the behavior of moving bodies, and could be applied to celestial phenomena as well as the movement of bodies on the surface of the earth. The theory of electromagnetism that had been developed by the Scottish physicist James Clerk Maxwell explained not only electrical and magnetic phenomena but also the behavior of light. Light, Maxwell said, was made up of oscillating electric and magnetic fields. He had pointed out that, if his theory was correct, other kinds of radiation could exist at frequencies far above and far below the frequencies that characterized light waves. This prediction had been confirmed in 1888 when the German physicist Heinrich Hertz discovered radio waves. He was able to show that these waves had both electric and magnetic fields. At first he thought they traveled at only two-thirds of the velocity of light. This error was soon corrected, and it was established that the waves propagated at light velocity, just as Maxwell had predicted.

In 1895, a number of prominent scientists still did not believe in the existence of atoms. Yet even they were willing to admit that the concept of the atom was at the very least a useful fiction not to be easily dispensed with, either in physics or in chemistry. And if one preferred to believe that matter was made up of these tiny particles, a plausible theory existed that seemed to describe what they might be like. In 1867, the Scottish physicist William Thomson had proposed a theory of vortex rings. In 1892, Thomson was to be raised to the peerage, becoming Lord Kelvin. During the Victorian era, England often honored its greatest scientists by knighting them or making them barons, a practice not observed in earlier times. Newton did not become Sir Isaac Newton because he was the most famous natural philosopher in the world (scientists were called "natural philosophers" then). He was knighted for his services as Master of the Mint.

According to Thomson, atoms were something like indestructible smoke rings. The properties of the chemical elements, he said, could be explained if it was assumed that they came in many different varieties. The sodium atom, for example, very probably consisted of two rings passing through each other like two links of a chain, and of course the vortex theory had no trouble explaining the emission of light from hot objects. If it was assumed the rings could vibrate, energy could be emitted or absorbed when the mode of vibration changed. A rapidly vibrating object, after all, contains more energy than one that is vibrating slowly.

In 1890, it seemed that there was little left to be done. Newton's laws explained the behavior of moving bodies. Maxwell's theory explained the phenomena of electricity, magnetism, and electromagnetic radiation. Finally, Thomson's theory explained the properties of atoms. In the eyes of many, physics was complete. To be sure, physicists were aware of a few unsolved problems here and there, but most scientists were sure that these little puzzles would eventually be cleared up. After all, the fundamental laws of nature were known. Scientists would continue applying them and, sooner or later, the science of physics would be complete.

THE BIRTH OF MODERN PHYSICS

Then, in the years 1895 through 1897, a whole series of revolutionary discoveries were made. The German physicist Wilhelm Roentgen discovered X rays in 1895. Just four months later, in 1896, the French physicist Antoine-Henri Becquerel discovered radioactivity. And in 1897, the English physicist J. J. Thomson* discovered the electron.

When Thomson made his discovery, British physicists had been theorizing about its possible existence for some time. Maxwell had suggested that "molecules of electricity" might exist. To us, the terminology sounds a little strange. After all, molecules are composite particles that are made up of atoms. But in Maxwell's day, the words *atom* and *molecule* were often used interchangeably. We can be fairly certain that "atoms of electricity" was what he really meant.

The idea was appealing. It made various electrical phenomena that were observed in the laboratory more understandable. Thus scientists sought to discover what properties the "atom of electricity" might have. In 1874, the Irish physicist George Johnstone Stoney attempted to estimate the electrical charge of the hypothetical particle. Stoney's estimate was off by a factor of almost twenty. In 1874, it was not known whether electricity was made up of particles with positive electrical charges, negative charges, or some combination of both. Nevertheless, the idea gained favor, and Stoney gave the electron its name in 1891.

Thomson discovered the electron some six years later. Though he could devise no way to measure either the mass or the charge of the particle, he was able to compute a ratio of mass to charge. He accomplished this in an ingenious manner. Thomson knew that charged particles were deflected both by magnetic and electric fields. The force exerted by an electric field depends only on a particle's electric charge. On the other hand, the force exerted by a magnetic field depends upon both the charge and the particle's velocity. So Thomson performed an experiment in which he adjusted the strength of the electric and magnetic fields until the forces they produced balanced one another out. This enabled him to calculate how rapidly the electrons were moving. He had two equations, one involving charge and one containing charge and velocity. When he put them together, charge canceled out.

Now, if Thomson had known how much an electron weighed, it would

---

* His full name was Joseph John Thomson. His contemporaries knew him as J. J., and he is still referred to in that manner today.

have been a simple matter to measure the particle's charge. All he would have had to do would be to turn off the magnetic field and measure an electron's deflection when it was influenced by the electric field alone. Unfortunately, he didn't know the electron's mass. Nevertheless, he could use this experience to measure the ratio of mass to charge.

Thomson found that this quantity was very small, which suggested that an electron probably weighed much less than an atom. Thomson's guess was correct. The electron has a mass that is very much smaller than that of the lightest atom, hydrogen. Like all atoms, hydrogen is electrically neutral. It consists of two particles, a positively charged proton and a negatively charged electron that orbits around it. But they do not contribute equally to the hydrogen atom's weight—the proton is 1,836 times heavier than the electron.

THE PLUM PUDDING ATOM

During the late 1890s, nothing was known about atomic structure. Hence it seemed reasonable at the time to assume that most atoms contained thousands of electrons. If a hydrogen atom contained 900, for example, that would account for half the mass. Approximately the same ratio would be maintained if oxygen contained 14,000 or 15,000. According to this scheme, an iron atom would have something like 50,000 electrons and an atom of gold approximately 180,000.

Two theories were proposed to explain the structure of the atom. The Japanese physicist Hantaro Nagaoka suggested that the negatively charged electrons formed rings that revolved around a positively charged sphere. In Nagaoka's model, atoms somewhat resembled the planet Saturn, whose rings are composed of small particles. But this theory possessed a grave drawback. According to Maxwell's electromagnetic theory, electrons that moved in circular or elliptical orbits would constantly emit radiation. If this were the case, the individual electrons in the rings would lose energy and fall into the sphere. Nagaoka's model seemed to produce atoms that would be unstable.

In the second theory, developed by Lord Kelvin (who had given up on his vortex-ring theory by this time; the discovery of the electron had made the idea untenable) and J. J. Thomson, the electrons were *inside* the sphere of positive electricity. The problem of energy loss still existed; the electrons would give off radiation wherever they were located. However, Thomson hoped it would be possible to show there could be configurations that caused this loss to be minimized. The Kelvin-Thomson theory was later to be called the "plum pudding model" of the atom. The positively charged sphere was the pudding, and the electrons were the plums.

By 1906, Thomson was able to deduce on theoretical grounds that the number of electrons in an atom was nothing like the number that had previously been assumed. He obtained this result by using three independent methods. The first made use of data concerning the interaction of light with transparent materials; the second concerned the interaction of X rays with gases; and the third the absorption of rapidly moving electrons in matter. In all three cases, Thomson assumed that the electrons in the atoms that constituted the matter played a prominent role.

All three methods led Thomson to the conclusion that atoms contained many fewer electrons than had previously been thought. In fact, the hydrogen atom probably had only one electron, while atoms of other elements possessed relatively small numbers. He was not able to state the exact numbers. According to the results he obtained, the number of electrons in an oxygen atom, for example, was somewhere between three and thirty-two. Oxygen atoms actually contain eight electrons. So Thomson's estimates were a great improvement over the previously accepted figure of approximately fifteen thousand.

THE DISCOVERY OF
THE ATOMIC NUCLEUS

Thomson's theory of the atom was ingenious, but it was soon contradicted by stunning new experimental results. In 1911, the British physicist Ernest Rutherford announced he had determined that electrons did not move about in a sphere of positive electricity. On the contrary, the positive charge—and most of the mass—of an atom was concentrated in a tiny nucleus in the atom's center.

The experiments that Rutherford performed involved bombarding gold foil, and then foils of other metals, with alpha particles emitted by a radioactive substance. He used foils because they could be made thin enough to allow most of the bombarding alpha particles to pass through. Gold in particular can be formed into sheets that are especially thin.

Alpha particles are composite particles that are made up of two protons and two neutrons. In 1911, the neutron had not yet been discovered, so the exact composition of alpha particles was not known. Their mass and charge could be measured, however, and it was determined that they carried a positive charge and were more than seven thousand times heavier than electrons.

Rutherford reasoned that if Thomson's theory was correct the alpha particles would, at most, experience small deflections when they encountered atoms of gold in the foil. Since alpha particles were so much heavier than electrons, the electrons would hardly affect the trajectories of the former. A collision between an alpha particle and an electron would be something like one between a battleship and a rowboat. Except that the electron would be deflected, not destroyed. On the other hand, the presence of spheres of positive charge with much larger masses would tend to push the alpha particles a little off course. Like charges repel one another, and both alpha particles and Thomson's spheres were positively charged.

To understand the nature of Rutherford's experiment, consider the following analogy: Suppose I fire a rifle at objects that stand on a fence a hundred yards away. One of them is a huge marshmallow about the size of a basketball. This corresponds to Thomson's atom.* The other is a steel ball about the size of a golf ball. If a rifle shot hits a marshmallow, it will pass right through. At best, it will be deflected slightly. On the other hand, if a shot hits a small steel ball, it will ricochet at some large angle. Since the ball is a very small object, I will probably fail to hit it most of the time. If I fire at it, most of my shots will not be deflected at all, and a few will glance off in random directions. Some will ricochet back in my direction.

---

* I use the example of a marshmallow because I want to ignore the minimal effects produced by the electrons. The plum pudding analogy just wouldn't work here. Electrons are very light and plums are not.

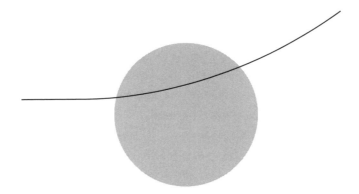

Figure 2a. When Rutherford began his experiments in which atoms were bombarded with alpha particles, it was believed that atoms were made up of spheres of positive electricity in which the electrons were embedded. He therefore expected that the alpha particles would, at most, be deflected at small angles. Here only the path of an alpha particle and the hypothetical positively charged sphere are depicted; the electrons—which were not expected to have any significant effect on the alpha particle—are not shown.

If an alpha particle collides with a gold atom, there will be no physical impact. But the effect will be similar. The alpha particle will be repelled by the positively charged nucleus. Furthermore, since a gold nucleus is almost fifty times as heavy as an alpha particle, it will be relatively little affected by the collision, while the alpha particle will be sent flying in one or another direction. Exactly what this direction is will depend upon how close the alpha particle's trajectory takes it to the atomic nucleus.

When Rutherford performed the experiment, he found that most of the alpha particles passed through the gold foil with little deflection, while a few were deflected at large angles. In some cases, these angles were larger than 90 degrees. In other words, a few of the alpha particles ricocheted backward. This implied that the positive charge in an atom was concentrated in a region that was much smaller than anyone had previously supposed.

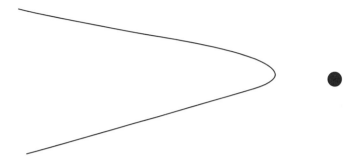

Figure 2b. When the experiments were performed, some of the alpha particles were deflected through large angles. The only possible way to explain this was to conclude that the positive charge in an atom was concentrated in a small region in its center. In this diagram, the positively charged alpha particle is strongly repelled by the positively charged atomic nucleus, and its change in direction is pronounced. Again, the electrons in the atom are not shown.

ATOMIC CATASTROPHE

The discovery that atoms contained tiny nuclei was a great experimental success, though it led to a theoretical catastrophe. Thomson had hoped that by using the plum pudding model, it would be possible to show that the amounts of energy that the electrons radiated would be minimal. But with Rutherford's nuclear model, this was impossible. The electrons had to be revolving around the nucleus in a manner similar to that in which the planets orbit the sun. Otherwise electrical attraction would cause them to fall into the nucleus. But Maxwell's laws said that electrons circling an atomic nucleus would radiate energy so rapidly that the atoms should quickly collapse. As radiating electrons lost energy, they would spiral in toward the nuclei. As they did, the energy radiated by any individual electron would rapidly increase. The closer it got to the nucleus, the stronger the attraction between the negatively charged electron and the positively charged proton would become. This growing force of attraction would cause the electron's velocity to increase. The faster the electron moved, the more energy it would emit. Electrons should spiral in toward nuclei and collide with them within a tiny fraction of a second. If both Rutherford's theory and classical physics were correct, there should be no atoms, no matter as we know it, and, of course, no physicists.

In fact, as British physicist and author Paul Davies has pointed out, the energy released under such circumstances should have been infinite. Davies begins by supposing that the electron and the proton are dimensionless point particles. This assumption is not as absurd as it sounds. As we shall see later, there are compelling theoretical reasons why the electron has to be viewed as a point charge. On the other hand, a finite size is attributed to the proton today because protons are now considered to be composite particles; they are made up of smaller particles called quarks. Yet as Davies points out, the existence of quarks was only first suggested about a half-century after Rutherford's experiment. In the early twentieth century the proton was not thought to be so unlike the electron. The only major difference between the particles seemed to be that the proton was much heavier and carried a positive charge.

If the electron and the proton are point charges, there is no limit to the amount of energy that can be released when a hydrogen atom collapses. As an electron lost energy, it would move closer and closer to the proton. As this happened, the electrical attraction of the proton would whip the electron around at higher and higher velocities. This is analogous to the steadily increasing speed of a comet as it draws nearer to the sun. As the electron begins to move faster, the rate of energy release will increase. When the electron and the proton finally collide, the rate of energy release will become infinite.

The physicists of the early twentieth century weren't worried about infinite quantities of energy. In their minds, the problem was one of explaining how atoms could exist. Davies' example shows how difficult the problem of atomic catastrophe was. According to the model that was commonly accepted in those days, the creation of a single atom should have created an explosion of infinite intensity.

ULTRAVIOLET CATASTROPHE

In order to see how atoms could exist and how that problem was eventually solved, we will backtrack a little and look at another infinity question that appeared around the turn of the century. This problem was related to attempts to find a mathematical formula that would accurately predict the amount of energy that a hot, glowing body would radiate away. When the solution was found, this led directly to the birth of quantum theory.

When constructing a theory, scientists generally begin by considering what happens under idealized conditions. Once the initial theoretical problems are solved, it is then possible to consider what happens in more complicated situations. The physicists considering this problem began by considering the case of a blackbody. A blackbody is an object that absorbs all the energy that falls upon it. If such a thing existed, it would appear perfectly black because it would reflect no light.

True blackbodies do not exist in nature. But it is not very difficult to construct apparatus that simulates the behavior of a blackbody in the laboratory. A hollow sphere that contains a small hole is such a device. Any light or other radiation that enters the hole will bounce around on the interior walls and eventually be absorbed. If the hole is small enough, very little light will make its way out again. And if the sphere is heated, its interior walls will begin to glow, and the light emitted from the hole will be like that which a theoretical blackbody would emit.

According to the theories accepted during the early years of the twentieth century, light was produced by vibrating electrons. Even though no one knew how atoms could be stable, it was still possible to calculate that if the electrons in an atom vibrated in a specific manner, then light or some other form of electromagnetic radiation would be produced. It seemed logical, therefore, that one should be able to use electromagnetic theory to calculate how the light emitted should vary with wavelength. The quantity of light emitted at different wavelengths is not the same. Different wavelengths correspond to different colors of light. A red-hot piece of iron is obviously emitting more "red wavelengths" than those of other colors.

Furthermore, there is a relationship between the wavelength pattern and temperature. As a piece of iron is heated, it will first retain its usual color because most of the energy it is emitting is given off in the form of invisible infrared "light." It will then take on a red glow, and as it is heated further, it will glow white. The English physicist Lord Rayleigh thought that it should be possible to describe all this mathematically, but when he tried to derive a mathematical formula on theoretical grounds, he obtained a result that couldn't possibly be correct.

THEORETICAL DIFFICULTIES

Rayleigh did find a mathematical expression (known as the Rayleigh-Jeans law; the English astronomer James Jeans is also given credit) that correctly predicted how much energy would be emitted at long wavelengths. In that region, the vibrating-electron model on which Rayleigh's calculations were based seemed to work. However, Rayleigh's calculation predicted that more and more energy would be emitted as the wavelengths grew shorter. In fact, it predicted that the quantity of energy emitted would be infinite. After all, there is no limit to how short the wavelengths of electromagnetic radiation can become.

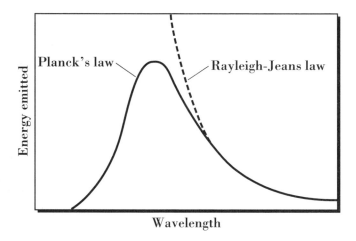

Figure 3. The ultraviolet catastrophe. A glowing object emits different amounts of energy at different wavelengths. The graph illustrates a typical example. Ultraviolet wavelengths are on the left and those in the infrared are on the right. The Rayleigh-Jeans law, which gave correct results for long, infrared wavelengths, predicted that energy emission would become infinite at short wavelengths. The graph also shows Planck's law, which is discussed on pages 84–85.

Short wavelength radiation carries larger amounts of energy than that of long wavelength. For example, ultraviolet radiation has higher energy than visible light. It is the ultraviolet component of solar radiation that causes tanning and sunburn. If we were exposed to only visible light when we went outside, we would never have to worry about staying out in the sun too long. X rays and gamma rays have even shorter wavelengths than ultraviolet "light" and are thus more powerful.

When physicists first began to struggle with the theoretical behavior of blackbodies during the 1890s, it had not yet been established that X rays were a form of electromagnetic radiation. Hence the prediction that energy emission should become infinite at short wavelengths was called the "ultraviolet catastrophe," a term attributed to the Austrian physicist Paul Ehrenfest.

The ultraviolet catastrophe, if it were a real effect, would be a catastrophe indeed. If energy emission did increase rapidly with shortening wavelengths, then every time an object was heated it would give off a blast of radiation that would vaporize the earth, the solar system, the rest of our Milky Way galaxy, and eventually the entire universe. Under such conditions, smoking would really be hazardous to your health. Anyone who struck a match to light a cigarette would destroy the universe.

The German physicist Wilhelm Wien performed studies of blackbody radiation in the laboratory and was able to find a mathematical formula that described the way in which the emission of light from a blackbody varied with temperature. This formula can also be found theoretically, using the principles of thermodynamics, the branch of physics that deals with heat and energy. But Wien's approach didn't work, either. His formula, known as Wien's law, gave the correct results at short wavelengths, but deviated from experimental observations at wavelengths that were longer.

So at the end of the nineteenth century, physicists had two theoretical formulas for the emission of radiation from blackbodies. The Rayleigh-Jeans law gave correct results at long wavelengths, but became infinite at short ones. Wien's law produced no infinities, but was incorrect at long wavelengths. Scientists found themselves in a quandary—they could not explain what appeared to be a simple, fundamental phenomenon, the emission of light from a hot body.

THE RELUCTANT REVOLUTIONARY

The solution to the blackbody radiation problem was found by the German physicist Max Planck in 1900. Planck began by looking for a mathematical formula that would correctly describe the behavior of blackbody radiation. After a great deal of mathematical tinkering, he found an expression that seemed consistent with experimental data. Then he asked himself what assumptions he had to make in order to derive this formula. To his surprise, he found that it was necessary to assume that light was emitted in tiny packets, or quanta, of energy. According to Planck's quantum theory, a hot body could throw off only whole numbers of energy quanta in any given period of time. It would emit one energy quantum, or two quanta, or 10 billion. But it could not emit one and a half quanta, or any other quantity that contained a fraction.

Planck was a conservative classical physicist, and he found it hard to accept the idea that blackbodies should emit light in the form of streams of energy packets. After all, it had been known for quite some time that light was a wave phenomenon. Waves simply were not created in this manner. The emission of light should be a continuous process, not one that involved little packets of energy. However, that indeed seemed to be the only solution that made sense mathematically. Years later, Planck was to describe his introduction of light quanta as an "act of desperation." He had been working on the problem of blackbody radiation for six years and he felt he had to find a solution "at any cost."

Most of the other physicists of the day disagreed with Planck's evaluation, and the quantum hypothesis received little attention during the next five years. In the meantime, Planck labored to find some more reasonable way to the correct formula. He had no way of knowing that the hypothesis of light quanta would turn out to be correct, and that in later years he would be considered the originator of the twentieth-century revolution in physics.

One physicist who did believe in Planck's quanta was Albert Einstein. In 1905, Einstein made the daring suggestion that light was actually propagated in the form of energy quanta. Light acted as though it was made up of particles. In making this suggestion, Einstein went much further than Planck, who had said nothing about the character of light as it traveled through space. Planck's theory required only that a hot body lose energy to its environment in this manner. He assumed that once light was emitted it traveled in the form of electromagnetic waves.

Of course, Einstein's hypothesis turned out to be absolutely correct. Today, we often speak of particles of light, which are called photons. Light does indeed have both a wave and a particle character. But in 1905, the idea seemed outlandish, even though it was capable of explaining various phenomena that were observed in the laboratory. The conservative Planck found it totally unacceptable. Nevertheless, Planck did admire the quality of much of Einstein's work, and recommended Einstein for membership in the Prussian Academy of Science in 1913. Planck's letter concluded with the words:

> That he may sometimes have missed the target in his specula-
> tions, as, for example, in his hypothesis of light-quanta cannot
> be held much against him, for it is not possible to introduce re-
> ally new ideas even in the most exact sciences without some-
> times taking a risk.

NIELS BOHR

In 1913, the same year that Planck wrote his now-famous letter to the Prussian Academy, the Danish physicist Niels Bohr announced his quantum theory of the hydrogen atom. Bohr showed that by making use of the concept of energy quanta it was possible to construct a theory that avoided the problem of atomic collapse. Although Bohr's theory proved not entirely satisfactory in the end, it was a revolutionary discovery. It was the first step toward an understanding of the behavior of matter on the subatomic level.

If Bohr had been more tactful when he was a young man, or if he had had a better understanding of English, he might never have achieved this result. Or at the very least, he might have created the theory somewhat later than he did. After receiving a Ph.D. from the University of Copenhagen in 1911, Bohr went to Cambridge where he wanted to work with

J. J. Thomson in the Cavendish Laboratory, an important research center at the time. Thomson was somewhat clumsy when it came to operating laboratory apparatus, and it was often said it was best not to allow him to get too near his own equipment. He was quite ingenious when it came to devising experimental methods, however, and it is significant that some seven of Thomson's research assistants later won Nobel Prizes for discoveries of their own. Collecting data for Thomson was excellent training.

Bohr's first meeting with Thomson did not go well. During the course of the meeting, Bohr picked up Thomson's book *Conduction of Electricity through Gases*, pointed to one of the mathematical formulas in it, and said, "This is wrong." After some other awkward encounters, Thomson began making detours to avoid meeting Bohr as he walked through his laboratory. For a while, Bohr kept busy, doing research and attending lectures. But then he decided to leave Cambridge and to go to Rutherford's laboratory at Manchester instead.

When he arrived at Manchester, Bohr was already familiar with Rutherford's nuclear model of the atom, yet he was not in a position to construct an atomic theory. He did not know how many electrons atoms of any given substance contained. Physicists no longer believed that atoms contained thousands of electrons, but no methods had been developed that would allow them to determine the exact number even in the simplest atoms.

One of the scientists working in Rutherford's laboratory was Charles Galton Darwin, the grandson of Charles Darwin. He had been working on some theoretical calculations concerning the energy lost by alpha particles when they passed through matter. When Bohr did some theoretical work of his own that elaborated upon that of Darwin's, he began to realize that the work done with alpha particles in Rutherford's laboratory strongly suggested that a hydrogen atom contained only one electron outside the positively charged nucleus.

Bohr didn't remain in Manchester long; he arrived in March of 1912 and left in late July. It had never been intended that his visit be anything but a short stay, but the knowledge he acquired during those four months provided one of the elements that made up his quantum theory of the hydrogen atom, which was published in 1913. Bohr could hardly have attempted to explain the behavior of this atom without knowing whether it contained one electron, or four electrons, or a dozen.

A THEORY OF THE ATOM

When Bohr propounded his theory of the hydrogen atom, he assumed that the electron revolved around the nucleus in a circular orbit. Next, he assumed that an electron occupying such an orbit would not constantly emit radiation. This was one of the most revolutionary proposals ever made in physics. For the first time, it had been suggested that particles in atoms did not behave the way that the known laws of physics said they would. Maxwell's theory of electromagnetism said that *any* charged particle revolving in a circular path should radiate energy. And indeed they do. In fact, nowadays electrons are often made to move at high velocities along circular paths in particle accelerators in order to produce what is called synchrotron radiation. If the electrons are moving fast enough, they will emit radiation in the high-energy X-ray region of the spectrum. These X rays can then be used to perform various kinds of experiments.

But the assumption that orbital electrons did not radiate was only the beginning. According to Bohr, only certain specific orbits were possible. Electrons could not occupy any positions between them. Electrons occupying orbits farther away from the nucleus possessed more energy than those in lower orbits, and atoms gave off radiation when an electron made a quantum jump from a higher orbit to a lower one, emitting one of Planck's quanta in the process. The reverse process could also take place. Electrons could also absorb quanta and move to higher orbits. Finally, there existed a lowest energy state, called the ground state. When an atom was in the ground state, the electron occupied its smallest possible orbit, and the atom could radiate no farther.

In Bohr's theory there were an infinite number of possible electron orbits,* and there was a finite distance between any two neighboring orbits. Thus the outermost orbit had an infinite radius and an infinite circumference. But, unlike the infinities encountered in radiative collapse of the

---

* No, this does not contradict the assumption that only certain specific orbits were possible. An infinite number of orbits is not the same as all possible orbits. An analogy should make this clear: The set of all numbers divisible by 10 is infinite, but this set clearly does not include all numbers.

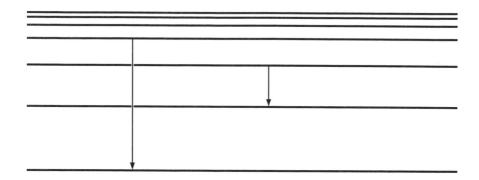

Figure 4. The energy levels of an electron in the Bohr atom can be depicted graphically. The spacing between the levels becomes smaller and smaller as the amount of energy increases, with an infinite number of possible levels. If an electron is given an amount of energy greater than the highest level, it is kicked out of the atom. The vertical lines in this diagram depict two possible quantum jumps from one level to another. Since both are downward jumps, the electron will give off a photon of light when it goes from one level to another.

atom or in the case of ultraviolet catastrophe, these were not troublesome. The existence of an infinite number of possibilities never causes problems. For example, if conditions had been a little different during the formation of the solar system, the earth might have an orbit that was more elongated than it is, or it might be closer to the sun, or farther away. The number of possible orbits is infinite. Similarly, even though there were an infinite number of possible orbits in Bohr's theory, the electron occupied only one at any given time. In any case, orbits for which the electron was very far away from the nucleus did not correspond to anything real. An electron that strayed too far away would eventually begin to interact with other atoms, and would no longer be attached to the original nucleus.

NEGATIVE REACTION

Bohr's theory successfully explained much about the behavior of hydrogen atoms. It correctly predicted the particular wavelengths of light seen when an electric current was passed through hydrogen gas. And it explained why atomic catastrophe did not take place; when an electron was in a ground state orbit, it could not get any nearer to the nucleus. Nevertheless, the theory was not immediately accepted. In fact, most reaction was negative. Not long after the theory was published, the German physicists Otto Stern and Max von Laue vowed to each other that if this crazy theory of Bohr's turned out to be correct, they would both leave physics. Of course, neither one did this. Von Laue went on to win the Nobel Prize for his work on X rays, and Stern later made important contributions to quantum theory himself.

As might be expected, Einstein's reaction was different. He commented that he had once had similar ideas himself but had lacked the courage to develop them. But Einstein was in the minority. The majority of the physics community was skeptical. One of the reasons for this skepticism was that Bohr's theory could explain the behavior only of the hydrogen atom. When attempts were made to apply it to the next simplest element, helium, it failed miserably. The theory seemed capable of explaining an atom that contained one electron, but not an atom with two.

Today we know why this was the case. After the theory of quantum mechanics—a more accurate successor to Bohr's theory—was discovered in 1926, it became apparent that electrons interacted with one another in ways that no one had previously suspected. Bohr's theory could not successfully describe the helium atom because no one understood how helium's two electrons really behaved.

Rutherford initially praised the theory, writing in 1914, "The theories of Bohr are of great interest and importance to all physicists as the first definitive attempt to construct simple atoms and molecules and explain their spectra."* Rutherford didn't actually endorse the theory. He simply said that while it was too early to tell whether or not the theory was true, it nevertheless had great importance and interest. This was uncharacter-

---

* The spectrum of an element is the pattern of light wavelengths that are seen when the element is heated to a high temperature.

istic behavior for Rutherford, who believed that the only real physics was experimental physics, that the "facts" were in the experiments. This attitude was often expressed in sarcasm. On one occasion, Wilhelm Wien, who was then the editor of *Annalen der Physik*, the journal that published Einstein's first papers on relativity, remarked to Rutherford that it seemed that no Anglo-Saxon could understand Einstein's theory. "No," Rutherford replied with laughter, "they have too much sense." Rutherford was a great experimenter, but he generally avoided theoretical issues. Perhaps it could be concluded that Bohr got along with Rutherford much better than he had with Thomson.

Rutherford's reaction to Bohr wasn't wholly favorable, however. Though he praised Bohr in print, he seems to have brought up an important objection in a private letter. Writing to Bohr in 1913, he said, "There seems to be one grave difficulty in your hypothesis. . . . It seems to me that you would have to assume that the electron knows beforehand where it is going to stop."

When he wrote this, Rutherford put his finger on an important point. An electron in one of the higher orbits could presumably jump down to a number of different states of lower energy. What determined which it would jump to? Bohr's theory didn't say. Nor did it say what caused an electron to execute a jump in the first place. Neither Bohr nor Rutherford could have known it, but the first step had been taken toward replacing the deterministic world of classical physics with the probabilistic world of quantum mechanics.

THE STRANGE WORLD
OF QUANTUM MECHANICS

Quantum mechanics was discovered by the German physicist Werner Heisenberg in 1925. Since quantum mechanics is the theory on which virtually all modern physics is based, its discovery has to be considered one of the most important scientific advances of the twentieth century. However, even though Heisenberg's theory attracted the interest of a number of leading physicists, little progress was made at first. Heisenberg had used what was then considered to be a strange kind of mathematics,* one with which most physicists were unfamiliar. Furthermore, the theory did not provide any intuitive picture of the behavior of subatomic particles. Physicists like to be able to create mental pictures of the objects they are dealing with, and Heisenberg's theory did not allow them to do that.

This situation didn't last for long. During the winter of 1925–1926, the Austrian physicist Erwin Schrödinger, using different starting assumptions, developed a theory called wave mechanics. Schrödinger worked out his theory during a vacation in the Swiss Alps. Leaving his wife at home, he took some scientific notebooks—and a girlfriend—to a villa he had rented. Dividing his time between physics and his companion, he soon produced a new theoretical description of the behavior of matter on the subatomic level.

In Schrödinger's theory, particles were pictured as waves that were spread out through space. Like light, matter was assumed to have both a wave and a particle character.** But it wasn't quite clear what exactly the waves were. They did not correspond to any physical quantities like the electrical and magnetic fields that constituted waves of light. However, it soon became clear that the theory yielded the correct results when it was used to perform calculations concerning the behavior of atoms.

For a short time it appeared that there were two new theories of the behavior of matter on the subatomic level, but it was soon discovered that Heisenberg's quantum mechanics and Schrödinger's wave mechanics were mathematically equivalent, even though they looked quite different. This seems to have pleased neither founder of quantum mechanics. Each intensely disliked the other's approach. During a conversation with the Austrian physicist Wolfgang Pauli in early 1926, Heisenberg remarked that Schrödinger's theory was "loathsome" and "crap." Schrödinger, on the other hand, charged that Heisenberg's method was "disgusting." Today both men are given credit for the discovery of quantum mechanics,

---

* Heisenberg made use of arrays of numbers called matrices. Today physicists use matrices all the time.

** This assumption was not original with Schrödinger. It had originally been proposed by the French physicist Louis de Broglie in 1924.

and Heisenberg's version is now called "matrix mechanics" to distinguish it from Schrödinger's.

By 1926 quantum mechanics had caught the interest of numerous physicists. It was soon discovered that quantum mechanics correctly described the behavior of the hydrogen atom, and a number of other problems in atomic physics were solved. However, one question remained: What, exactly, *were* Schrödinger's mysterious waves?

That problem was solved in 1926, when the German physicist Max Born showed that they could be interpreted as waves of probability. They had a character entirely unlike such things as light waves or ocean waves. If one was considering an electron, for example, it was most likely to be in places where the wave amplitude was highest, and least likely to be near places where this quantity fell to zero.* For the first time, physicists were working with a theory in which they could not assign a definite location in space to an object.

One man who was less than delighted with this finding was Schrödinger. Like Planck, he was something of a scientific conservative. He had hoped that by using a wave formulation a way could be found to avoid quantum jumps, and a classical picture of the atom could be restored. On one occasion, after a series of conversations with Bohr, Schrödinger exclaimed, "If you have to have these quantum jumps then I wish I'd never started working on atomic theory!"

CLASSICAL PHYSICS VS. QUANTUM PHYSICS

Quantum mechanics was unlike any theory that physicists had encountered before. The difference can be illustrated by comparing the behavior of a large, macroscopic body to that of a very small one. For example, the position in space that will be occupied by the earth at any given time can be calculated with very great accuracy. Similarly, if we knew all the

---

* The probability was proportional not to the amplitude, or height, of the wave but to the square (technically the "absolute square") of the amplitude. One can find an analog to this in ocean waves. A wave that is higher will also be broader, and the amount of water it contains will be roughly proportional to the square of its height.

forces acting on a baseball, we could calculate exactly its path as it flew through the air. In quantum mechanics, nothing like this is possible. At any time, there will be a certain probability that an electron in a hydrogen atom, for example, is in one place, and there will also be a certain probability that it is in another.

If an electron in an atom is a bundle of probabilities, this has the effect of "smearing it out" over space. Indeed, physicists often speak of the "electron cloud" that surrounds an atom. In quantum mechanics, electrons no longer have orbits. Instead, they form probability clouds of different sizes and shapes. In the jargon of physics, these configurations are known as quantum *states*. When a quantum jump takes place, the electron does not go from one orbit to another; rather, it makes a transition between two different states. Since different states are associated with different energies, quantum jumps still produce the emission of light.

Human language evolved to describe events in the macroscopic everyday world, and that language often proves inadequate for describing happenings on the subatomic level, where objects behave differently. Nevertheless, an electron in a hydrogen atom can be viewed, in some sense, as being in an infinite number of different places at the same time. The probability waves surround the nucleus, after all, and space contains an infinite number of points.

This is another example of a kind of infinity that is quite benign. No infinite quantities—such as infinite energies—appear. Like a planet circling a sun, an electron can have an infinite number of different positions. The only difference is that it occupies them all simultaneously.

Quantum mechanics is full of probabilities, and the electron cloud is only one example. The theory does not tell us when an electron will make a transition from one state to another; it only gives us a probability. Furthermore, one can speak of the probability of a subatomic particle being in this or that state. In fact, subatomic particles are generally considered to be in mixtures of states. Not only is an electron many different places at once, it can simultaneously occupy an infinite number of different energy states.

You might think it impossible to perform calculations with such a theory, but this is not the case. It is true that the theory does not tell us what an individual particle will do. However, matter is made up of very large numbers of particles. For example, there are approximately 10^{23} (the number represented by the numeral "1" followed by twenty-three zeros) hydrogen and oxygen atoms in one cubic centimeter of water. And when quantum probabilities are averaged out over large number of objects, very accurate predictions can be made. Furthermore, quantum mechanics does allow certain quantities, such as energies and light wavelengths, to be calculated exactly. The validity of quantum mechanics has been confirmed over and over again in numerous experiments of high precision. There can be no doubt that it is a correct theory.

6

ELECTRONS HAVE INFINITE MASS

QUANTUM MECHANICS has the potential to provide a fundamental explanation for almost everything. It not only explains the structure of atoms, it can also be used to determine how atoms combine to form molecules. It has become the central theory of both physics and chemistry. The properties of all materials depend upon how their atoms and molecules interact with one another. Thus quantum mechanics provides the ultimate explanation for the strength of steel, the burning of gasoline, the springiness of springs, the softness of pencil lead, and the processes that take place in the circuitry of a computer chip. Quantum mechanics is the basis for theories of the behavior of subatomic particles and is used in theories that attempt to explain the origin of the universe. The only fundamental phenomenon that quantum mechanics does not help to explain is gravity.

However, before quantum mechanics could become a theory of almost everything, two important gaps had to be bridged. The theory that was developed by Heisenberg and Schrödinger in 1925 and 1926 did not explain the interaction of electromagnetic radiation with matter, nor was it fully consistent with Einstein's special theory of relativity. According to the theory of quantum mechanics, a photon (the modern term for a quantum of light) was emitted when an electron in an atom jumped from one energy state to another. Quantum mechanics said nothing about how this happened, however. Nor did it explain the reverse of this process, the absorption of a photon by an atom.

The two problems that physicists faced in the late 1920s were clearly connected. Quantum mechanics had to be combined with relativity if one was to have any chance of accurately explaining the interaction of light and matter. The theory of relativity deals with the behavior of objects that travel at velocities close to the speed of light. There must be relativistic effects when photons interact with atoms. For example, when light is reflected from a mirror, a photon that had been traveling at the speed of light in one direction suddenly finds itself moving at the same velocity in the opposite direction. Something dramatic must have happened to the electron or electrons that caught it, turned it around, and sent it on its way. If indeed that was what really happened—the original theory of quantum mechanics did not say whether the atoms of a mirror reflected photons or whether they absorbed them and then reemitted them in the opposite direction. Furthermore, no one knew how an atom was able to do either of these things.

MISS DENT IS DETERMINED

The man who solved both of these problems, the English physicist P. A. M. Dirac,* was one of the greatest physicists of the twentieth century. If not for a series of chance events, Dirac might never have become a physicist at all. Although he had a strong interest in mathematics as an adolescent, he decided to study electrical engineering when he entered England's Bristol University. An engineering degree would offer a much better chance of making a living than one in mathematics, so when Dirac graduated in 1921, he immediately began to look for an engineering job. But England's economy was in recession at the time, and he could not find work. Dirac returned to Bristol to do postgraduate work, this time in mathematics. At the beginning of his second year of graduate study, he

---

* His full name was Paul Adrien Maurice Dirac, but he always signed his letters and scientific papers with the initials P.A.M. Even his closest colleagues had to guess what those letters stood for.

had to make a choice, whether to specialize in pure or applied mathematics. As it happened, there was only one other advanced mathematics student at Bristol at the time, a Miss Dent (first name unknown), who was determined to study applied mathematics. Dirac, who found himself indifferent about the matter, decided to study applied mathematics, too. He didn't want to make the professors teach two separate sets of courses for just two students.

In the British universities of the day, theoretical physics was considered part of applied mathematics. Most of the university physics departments in England had only experimental physicists on their faculty. The theoretical physicists, on the other hand, were generally found in the mathematics departments. A vestige of this tradition remains today, when Stephen Hawking is Lucasian Professor of Mathematics at Cambridge University, not a professor of physics. The Lucasian chair, incidentally, was previously occupied by both Newton and Dirac.

Dirac obtained his Ph.D. at Cambridge in 1926. By the end of the year, he was already making important contributions to physics, writing a paper that laid the foundations of quantum field theory, the theory that describes the interaction of light and other kinds of radiation with matter. Dirac was not solely responsible for creating the theory. Important contributions were also made by Heisenberg, by Pauli, by the German physicist Pascual Jordan, and others. And, as we shall see, the theory did not attain its final form until 1947 when the American physicists Richard Feynman and Julian Schwinger and the Japanese physicist Shin'ichiro Tomonaga showed how certain theoretical difficulties could be removed, an achievement for which they were jointly awarded the Nobel prize.

COMBINING QUANTUM MECHANICS AND RELATIVITY

Dirac knew that a successful quantum field theory had to incorporate the special theory of relativity, but this was not an easy problem to solve. Schrödinger had attempted to incorporate relativity into quantum mechanics in his original paper on the subject, in 1926, and had failed. Other prominent physicists also worked on the problem, but they were not successful, either. It appeared that putting quantum mechanics and relativity together obtained results that were not in agreement with experiments.

When theoretical physicists attempt to create new theories, they generally begin by trying to form mental pictures. A scientist who is adept at doing this is often said to have good "physical intuition." The next step consists of writing down mathematical equations that correspond to this mental picture. In his search for a relativistic quantum theory, Dirac followed a different procedure. He played around with mathematical equations until he discovered something that looked as if it might describe the subatomic world in a relativistic manner. This wasn't a random approach. Dirac discovered his theory because he was able to recognize a significant mathematical result when it appeared, and to develop it further.

Since Dirac's work was so highly mathematical, it is impossible to describe the steps by which he came to his theory in a book of this nature. It's enough to know that once the theory was completed, it gave a physical picture of the behavior of electrons and of atoms. As we shall soon see, this picture turned out to be a very strange one.

Dirac published his theory in 1928. Although his original intention was to try to understand the interactions between light and matter, his theory was also an improvement on the Heisenberg-Schrödinger quantum mechanics. It was capable of describing all the phenomena the older theory explained, often with greater accuracy, and could be used to obtain results that had previously been intractable. However, a problem soon appeared. Dirac's theory seemed to imply that electrons could have states of negative energy, and it wasn't clear how this should be interpreted. Furthermore, if you used Einstein's equation $E = mc^2$ (here E stands for energy, m for mass, and c for the speed of light) to describe these electrons, it appeared you would have to conclude that they also had negative mass. If E was negative, m had to be negative also, because the term c^2 was always positive.

Negative quantities appear in mathematics and physics all the time. In many cases they can simply be ignored. For example, suppose you are told that a square has an area of 4 square inches and are asked to find the length of one side. All we have to do is to take the square root of 4, which is 2. Now it so happens that the number –2 is also a square root of 4. When the multiplication –2 × –2 = 4 is carried out, the minus signs cancel. But it does not follow that the square can also have sides that are –2 inches in length. We simply ignore the extra, spurious solution.

This cannot be done in quantum mechanics. The negative energy levels could not be ignored, because quantum mechanics allows electrons to jump between states of different energy, and there was nothing to prevent a particle from jumping from a positive to a negative state. It appeared that if Dirac's theory were correct, all of the electrons in the universe would eventually wind up with negative energy. And there was no stopping there. An electron in a negative energy state could then drop to any of an infinite number of states with even less energy, emitting radiation as it did. In the end, infinite quantities of radiation would be released, and every electron would wind up with infinite negative energy.

AN INFINITE SEA OF ELECTRONS

Dirac realized that some way had to be found to explain why positive-energy electrons existed at all. He soon hit upon a solution that was bizarre, but workable. The reason why positive-energy electrons were seen was that all of the infinite number of negative-energy levels were filled. There existed an infinite sea of negative-energy electrons.

Apparently, these negative-energy electrons could not be seen. However, Dirac realized that his idea had observable consequences. If one of the negative-energy electrons acquired enough energy, it would make a quantum jump into the positive-energy region, and an ordinary electron would suddenly seem to pop into existence. At the same time, an empty spot—a hole—would appear in the negative-energy sea. This hole would have the appearance of a positively charged particle.

In order to see what Dirac meant, consider a line of billiard balls:

Now imagine that one of the balls is removed:

Imagine next that the balls on the right-hand side of the figure move one by one to the left:

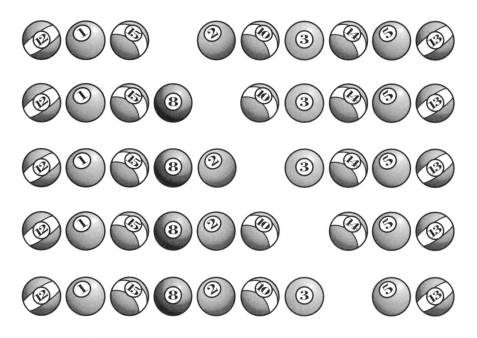

As the figures clearly show, the gap between the balls seems to be moving to the right.

Dirac realized that if you could not see the negative-energy sea, a hole in that sea would nevertheless be observable. If such holes existed, they could move in the way that ordinary particles did. Furthermore, they would have the appearance of positively charged particles with ordinary, positive energy. The energy that the negative electron absorbed would produce two particles, one with a negative and one with a positive charge.

At the time, physicists knew of only three particles: the proton, the electron, and the photon. The neutron would not be discovered until 1932. So at first Dirac wondered if the positively charged holes might not be protons. On further thought, he concluded this was not very logical. After all, a proton was almost two thousand times as heavy as an electron. So in 1931 he published a paper in which he suggested that a hole in the negative-energy sea would be an entirely new kind of particle. It would have the same mass as the electron, but it would have a positive, rather than a negative, electrical charge.

THE DISCOVERY OF THE POSITRON

Dirac's positively charged electrons were discovered the very next year by the American physicist Carl Anderson. Anderson was studying cosmic rays, which are not a form of radiation, but rather streams of energetic particles—mostly protons—that fall on the earth from space. Anderson observed them by using a cloud chamber. When particles entered the chamber, they would leave tracks, so that it was possible to study their trajectories. As the particles passed through the chamber, they were subjected to magnetic fields. This allowed Anderson to determine both the particles' charge and their mass. Magnetic fields cause charged particles to follow curved paths. A particle with a positive charge will veer off in one direction, and a negatively charged particle in the other. Furthermore, the path of a relatively light particle will curve more strongly than the path of a heavy one, which possesses more inertia, and which therefore cannot be pulled off course as easily.

Anderson's experiments revealed the existence of a particle with the same mass as the electron, but an opposite charge. Its path curved the same way an electron's did, but in the opposite direction. The new particle, which was soon named the positron, was exactly what Dirac had predicted.

This was the first time that the existence of a particle had been predicted on theoretical grounds before it was seen in the laboratory. Both Anderson (who was later to receive the Nobel Prize for his work) and Dirac received wide acclaim. One of the few negative reactions was from Rutherford, who was not amused. He remarked that it would have been more to his liking if the experimentalists had been allowed to discover the positron and determine its properties before it became part of theory.

Physicists generally considered the discovery of the positron to be a confirmation of Dirac's theory. However, many of them remained uneasy about the idea of a sea of negative-energy electrons. Not only did such a sea have to contain an infinite number of particles, it also had to possess infinite negative energy. Then, in 1934, the American physicist Robert Oppenheimer and Robert Furry, a postdoctoral fellow, put these concerns to rest by proving that it was possible to forget about the infinite sea and simply talk about electrons and positrons instead. This could be accomplished by making a simple change in Dirac's mathematical notation.

It may seem surprising that Dirac's theory could give rise to two entirely different pictures of the subatomic world. But remember that we have already encountered one example of this. The Heisenberg and Schrödinger formulations of quantum mechanics also seemed quite different, but they later turned out to be mathematically equivalent. The Schrödinger theory was based on the concept of waves, while the Heisenberg version said nothing about waves at all.

It is often difficult to form a precise mental picture of subatomic phenomena. There is no reason to expect that events in the subatomic world should precisely resemble those in the world of everyday experience. We can easily visualize the behavior of waterfalls, of cats and rainstorms because they are things that we have seen many times. A particle like an electron, on the other hand, behaves in unfamiliar ways, none of which correspond to anything in our visual experience. Sometimes many different mental pictures are possible.

If two different pictures lead to the same mathematical results, then they are equivalent. The reason that one can either think of "holes" in an infinite sea or "just electrons and positrons" is that both conceptions lead to the same results. As a matter of fact, there is yet a third way of thinking of positrons. As Richard Feynman has shown, a positron is mathematically equivalent to an electron that is moving backward in time. In this case, it is not necessary to believe that anything really moves from the future into the past, although it is sometimes convenient to imagine that it does. Under some circumstances this makes it easier to understand what is going on in the microworld.

PARTICLES, ANTIPARTICLES, AND VIRTUAL PARTICLES

According to Dirac's theory, electrons and positrons are created together in pairs. This is a process that has been observed many times. Under certain circumstances, a gamma ray will be transformed into an electron and a positron. When this happens, the positron does not continue to exist for very long. As soon as it encounters another electron, the two particles will annihilate each other and a second gamma ray will appear in their place. The energy of these gamma rays can be computed using Einstein's formula $E = mc^2$. Here, m is twice the mass of the electron, since two particles with identical masses are created.

The positron is said to be the antiparticle of the electron. Indeed, every known particle has an antiparticle. There are antiprotons, antineutrons, and antiquarks, for example. Strictly speaking, the positron should be referred to as an "antielectron." However, the positron received its name before the concept of antiparticles was developed, and the name has stuck.

Particles and antiparticles can be created out of pure energy. Surprisingly, they can also be created when sufficient energy is not available. This is a consequence of one of the fundamental ideas of quantum mechanics, Heisenberg's uncertainty principle. According to this principle, it is not possible to measure the position and momentum of a particle at the same time. The better one quantity is known, the more uncertainty there is in our knowledge of the other. The uncertainty does not arise from any limitations in our measuring instruments. In quantum mechanics a particle does not have an exact position or an exact momentum. This is why the electrons in an atom must be thought of as something like a cloud that surrounds the nucleus.

Now it so happens that Heisenberg's principle also applies to many other pairs of quantities. One such pair is energy and time. For example, the more accurately we know a particle's energy, the less able we are to predict how long it will remain in that energy state. The reverse is also true. If we knew very precisely the length of time that a particle remained in a state, the amount of energy that it had would be very uncertain.

One can go even further. When we deal with times that are very short, energy uncertainties become so large that particle-antiparticle pairs can be created. These virtual particles pop into existence out of nothing and then disappear again in about one-billionth of one-trillionth of a second. This is not just a theoretical idea. The creation of such virtual particles is a real phenomenon. Although such short-lived particles cannot be observed, their creation and annihilation has experimental consequences. The relevant experiments have been performed and the existence of virtual particles has been confirmed many times.

INFINITELY SMALL ELECTRONS
HAVE INFINITE MASS

Electrons are the particles that emit or absorb photons of light. A theory that attempts to describe the interaction of light with matter needs to deal with only three particles: electrons, photons, and positrons (the reason for the inclusion of this last particle will soon become clear). The quantum field theory that describes such interactions is known as quantum electrodynamics, or QED. Although QED has an imposing name, the basic idea is a simple one: Electrons and photons interact with one another in an environment in which virtual particles—virtual photons as well as virtual electrons and positrons—are constantly being created and destroyed.

Before I describe the effects of virtual particle creation, it will be necessary to look a little further at the properties of electrons. During the years immediately following the discovery of the electron, this particle was considered to be a tiny sphere. Physicists even tried to calculate the electron's size, using theoretical principles, but they soon realized that such attempts were doomed to failure and that the electron had to be viewed as a dimensionless point particle.

According to Einstein's theory of relativity,* no object can travel faster than the speed of light. Furthermore, no physical influence or signal can travel faster than light. This has profound consequences for physicists' conceptions of the electron and other subatomic particles. In order to see why, imagine that an electron experiences a collision with another particle. If the electron has a finite size, the force of the collision will be transmitted across the electron at a speed no greater than that of light. Thus the electron will become deformed for an instant.

This is exactly what happens when macroscopic objects collide. For example, when one billiard ball strikes another, both will become slightly deformed at the point of impact, then they will spring back to their former shapes. This causes no conceptual problems. Indeed, the idea helps us to understand the forces that the balls exert on one another. But in the case of an electron, matters are different. If an electron could be deformed, it could presumably be broken apart in a very high-energy collision. But this does not happen; all the experimental evidence we have indicates that the electron is an elementary particle. Furthermore, a sphere that can be deformed and which will then regain its former shape must have internal forces that produce this kind of elasticity. But there doesn't seem to be any sensible way of describing what such forces would be like.

The idea that the electron is a point particle has been confirmed to a certain extent by experiment. The reason I say "to a certain extent" is that only an experiment of infinite accuracy could confirm that the dimensions of an electron were exactly equal to zero. However, it has been shown that the electron's radius is less than a billionth of a billionth of a centimeter. Another way to express this would be to say that its maximum radius is at least 100,000 times smaller than that of a proton, and that its volume is at least 10^{15} (1 million billion) times smaller. Since these figures are upper limits, they are entirely in accord with the idea of zero dimensionality.

---

* When I refer to the theory of relativity, I mean Einstein's special theory of relativity, which was published in 1905, and not the general theory (Einstein's theory of gravitation) of 1915. This is common usage.

However, if an electron is a dimensionless point, then it must have infinite mass. In order to see why this is so, consider the case of two point charges. If they are like charges—both positive or both negative—they will repel each other. The force of repulsion follows an inverse square law. Thus if we want to force them closer together, an expenditure of energy is required. The closer they are to each other, the stronger the force of repulsion becomes. If they were moved so close together that there was zero distance between them, the force of repulsion would become infinite. In order to accomplish this, an infinite expenditure of energy would be required.

Something analogous happens in the case of the electron. It is possible to perform a thought experiment in which one imagines the electron to be a tiny sphere, which is then shrunk to a point. If one imagines the electron to be made up of a lot of little pieces of electrical charge, then these little charges move closer and closer together as the electron becomes smaller. When the electron's dimensions decrease to zero, there is zero distance between these charges, and the energy of the electron becomes infinite.

But as Einstein showed, mass and energy are equivalent. Thus the electron must have infinite mass. Of course, no one believes this is literally true. But it happens to be true in the best theory that we have at the moment. And, until a better theory comes along, viewing the electron in this manner will continue to be unavoidable.

VIRTUAL PARTICLES TO THE RESCUE

According to QED, we never see these "bare" infinite-mass electrons. We can only view them through the swarms of virtual particles that surround them. These virtual particles will not exist for very long, but others will constantly be created to take their place and the electron will be surrounded by a cloud of virtual particles. Furthermore, this cloud will be polarized. The positively charged positrons will tend to be pulled toward the electron, while the negatively charged virtual electrons will be repelled. There will be more positive charge near the electron, and more negative charge farther away from it. The "shielding" created by this polarized cloud has the effect of preventing us from observing the electron's infinite mass, or its infinite charge. Thus the quantities that are actually observed in the laboratory are finite. Yes, in QED, the electron has an infinite "bare" charge, too. And the virtual particles that surround it are infinite in number.

At first sight it appears that if a prize were awarded for the craziest theory ever conceived by physicists, QED would have little competition. Indeed, some very strange things happened when physicists first tried to use the theory to make calculations. Since the theory was mathematically complicated, and its equations could not be solved exactly, the physicists did what they always do in such cases. They solved the equations step by step. First, they looked for answers that were approximately correct. Then they used these to compute corrections, sure that they would get more accurate results. Normally, this procedure works quite well. It can be carried out as many times as any physicist would want, and the results will become more accurate each time.

But this didn't happen in the case of QED. The physicists who worked with it discovered that they could obtain results that were roughly correct if they calculated only to the first approximation. But if they attempted to obtain more exact results by calculating further, their equations "blew up," and infinite quantities appeared. This was both heartening and frustrating. The fact that approximately correct results could be obtained indicated that there was some element of truth in this odd theory, while the appearance of infinite quantities meant that there was also something seriously wrong with it.

During the 1930s, scientists used QED to obtain a number of important results. It appeared that the theory could indeed be applied to a number of different processes that took place on the subatomic level, if answers that were only approximately correct were satisfactory. However, few attempts were made to refine the theory so that more exact calculations could be made. In order to do that, it would have been necessary to somehow get rid of the infinities. And no one knew how that could be done.

RENORMALIZATION

Physicists often make their most important contributions while they are still in their twenties. Newton discovered his law of gravitation before he was twenty-five; Einstein was twenty-six when he published his first papers on the theory of relativity; Heisenberg was twenty-three when he discovered quantum mechanics. But Julian Schwinger was more precocious than any of them. Schwinger entered New York's City College at the age of fourteen. At sixteen, he wrote his first paper on QED. He didn't publish the paper. He did write a Ph.D. thesis during the summer that followed his graduation from college, but he wasn't awarded the doctorate immediately. It seemed that Columbia University (where he did graduate study) had a residency requirement and didn't immediately give Ph.D.'s to people who had just obtained bachelor's degrees. Schwinger had to wait until he was twenty-one before he was awarded the doctorate. After that, his career advanced rapidly, and he became a full professor at Harvard before he reached the age of thirty.

Given Schwinger's early interest in QED, it is not surprising that he should have found a way to deal with the theory's infinity problems. The technique he invented is called renormalization. Basically, renormalization is a way of subtracting the offending infinities so that only finite results are obtained.

Sometimes ideas in physics can be verbally described in simple ways and yet be mathematically complex. An example is the idea of curved space that is associated with Einstein's general theory of relativity (which I will discuss in a subsequent chapter). This simple description obscures the fact that Einstein's mathematical equations can become quite complex when an attempt is made to use them to solve specific problems. Indeed, there are instances where they become so complex they have never been solved. Similarly, renormalization is a mathematical technique of some complexity, as witnessed by the fact that the problem of removing the infinities from QED baffled an entire generation of physicists.

Schwinger announced his discovery of renormalization at a conference held at Shelter Island, New York, in 1948. When he did, he discovered he was not the only person to discover how to remove the infinities from QED. His talk was followed by one given by Richard Feynman, who also demonstrated how this could be done.

The two presentations could not have been more different. Schwinger's method was mathematically difficult, but it made use of a kind of mathematics that was familiar to the physicists in the audience. If they didn't follow all the details, they still knew that they would have no trouble understanding renormalization when they studied the technique in more detail later. Feynman, on the other hand, drew a lot of little pictures, used unfamiliar mathematical tricks, and talked about a new kind of mathematics that he had invented to express his ideas. The scientists sat there listening, baffled. At one point, Bohr, who was in the audience, made some remarks indicating that he wasn't sure whether Feynman even understood the basic principles of quantum mechanics.

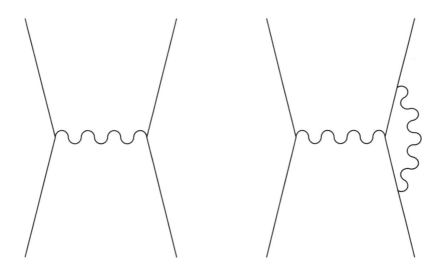

Figure 6. Two simple Feynman diagrams. In the diagram on the left, two electrons are moving in an upward direction. One emits a photon, which is absorbed by the other. The wavy line represents the photon. The diagram on the right shows a similar interaction. Here, one of the electrons emits a second photon, and then reabsorbs it. Feynman diagrams can become quite complicated because many different interactions can take place.

After the meeting was over, Schwinger and Feynman compared notes. Neither really understood the other's method. However, they discovered that when they did certain specific calculations, they got the same answers. Perhaps their mathematics was different, but their methods contained the same physics.

In 1949, the English-American physicist Freeman Dyson proved that Schwinger's and Feynman's methods were mathematically equivalent, and they share the credit for the discovery of renormalization with the Japanese physicist Shin'ichiro Tomonaga, who discovered Schwinger's method independently. Today, physicists who work with QED almost always use Feynman's method because it is mathematically less difficult than Schwinger's, and because Feynman's little pictures (now called Feynman diagrams) provide an intuitive picture of the processes taking place.

QED is a highly mathematical theory, and renormalization was a technique designed to eliminate a mathematical problem, the appearance of infinite quantities. Nevertheless, it is possible to say something about its physical meaning. Recall that QED pictures the electron as a particle with infinite charge and infinite mass that is shielded by a cloud of virtual particles. In effect, renormalization is a way of adding up the contributions of the virtual particles in such a way as to cancel these infinities out.

A HORRIBLE SCRAPING NOISE

Was renormalization really a logical mathematical procedure? Dirac didn't think so. He saw it as a way of avoiding fundamental problems, and he continued to be critical of the method until he died. "This is just not sensible mathematics," he said. "Sensible mathematics involves neglecting a quantity when it is small—not neglecting it because it is infinitely great and you do not want it!" Even Feynman agreed to some extent, saying that renormalization "is what I would call a dippy process! Having to resort to such hocus-pocus has prevented us from proving that the theory of quantum electrodynamics is mathematically self-consistent. . . . I suspect that renormalization is not mathematically legitimate."

Harvard University physicist Howard Georgi has used even more colorful language to describe QED and the renormalization process. Georgi compares QED to a car that "makes a horrible scraping noise and, after a minute or two, starts smoking and smelling awful and stops running." But if the theory is renormalized, as Georgi puts it, "I gun the engine and pound on the dashboard, the scraping sound goes away and the car runs beautifully."

As Georgi implies, renormalization has made QED a remarkably successful theory. QED has yielded predictions that have been experimentally verified to an accuracy of better than one part in ten billion. Furthermore, it explains not only the interaction of light with matter, but also the nature of electric and magnetic forces. According to QED, charged particles interact with one another by exchanging photons. Each particle emits photons that are absorbed by others. Such interactions can lead to either attractive or repulsive forces. It is the exchange of photons that causes two electrons to repel one another, and it is the exchange of photons that is responsible for the force of attraction between an electron and a proton.

This can be illustrated by an analogy. Suppose two skaters throw a heavy ball back and forth. When the first skater throws the ball, he will recoil a bit. When the second skater catches it, he will be pushed slightly backward. Throwing the ball back and forth will create a kind of repulsive force. Next, suppose that the skaters have their backs to each other and are throwing a boomerang. When the first throws the boomerang, he is nudged toward the second. If the boomerang flies in a large circle so that it is caught by the second skater, the second skater will be pushed toward the first. Admittedly, this example is a little far-fetched, but it does show how an attractive force would be produced.

STRONG AND WEAK FORCES

The list of theoretical successes does not end here. Other renormalizable quantum field theories modeled on QED exist that have also proved to be extremely successful. For example, a theory called quantum chromodynamics, or QCD, explains the structure of such particles as protons and neutrons and the forces that bind them together in atomic nuclei.

Protons and neutrons are made up of quarks. A proton contains one down quark and two up quarks, while a neutron contains two down and one up quark. Here, the words *up* and *down* should not be taken literally. They are just names. There are actually six quarks in all, but the other four, the strange, charm, top, and bottom quarks, are not constituents of ordinary matter; they make up particles that are seen only in the laboratory.

Quarks carry a kind of charge that is analogous to positive and negative electrical charges. But there is an important difference. Quark "charges" come in three varieties rather than two. They are called red, green, and blue, after the three primary colors of light. Again, do not take the names literally. Quarks have no color in the ordinary sense of the term; this would be impossible, because they are much smaller than the wavelengths of light. If one named the charges plus, minus, and extra, or even Hans, Gretchen, and Amy Jo, nothing would be changed.

Just as positively and negatively charged particles interact with one another by exchanging photons, quarks are bound together because they exchange particles called gluons. The force created by gluons gives rise to the so-called strong force, which binds protons and neutrons together in nuclei. The strong force is quite powerful; it easily overcomes the repulsion that exists between pairs of positively charged protons. But its strength falls off rapidly to zero with distance. Thus it is a force that operates only within the nucleus.

In nature there are four forces: the strong force, the weak force, the electromagnetic force, and gravity. The weak force also operates only within the nucleus; it is responsible for certain types of nuclear reactions, such as beta decay (the emission of an electron from a nucleus). Unlike the strong and electromagnetic forces, the weak force cannot be renormalized. However, a quantum field theory that combines the weak force with the electromagnetic force can be. The theory of this "electroweak" force predicts the existence of new particles, called the W and Z particles (discovered in 1983).

The force of gravity cannot be renormalized. However, we must conclude that QED and theories modeled on it provide a fundamental explanation for all natural phenomena except gravity. Yet there are drawbacks. For example, QED does not tell us what the mass and the charge of the electron should be. It is necessary to "plug in" the observed values of these quantities. And of course physicists would like to be able to show that the theories they use are mathematically consistent. But, as Feynman remarked, this probably cannot be done. Most likely, subtracting out infinities is an illegitimate procedure.

EXPLAINING EVERYTHING

Physicists have long sought a theory that would explain all the forces of nature. In other words, they would like to discover an analog of the electroweak theory that combines all four forces, not just two. Such a theory would have great aesthetic appeal; it is always nice to be able to tie up everything in a neat little package. Furthermore, such a theory would likely predict new natural phenomena, such as the W and Z particles that appeared when the electromagnetic and weak forces were combined. Finally, scientists suspect that if such a unified theory were found, it would predict the charges and masses of observed particles. We would know, for example, why a proton is 1,836 times as heavy as an electron.

Direct attempts to create such a theory have not been very successful. A number of theories attempt to combine the strong, weak, and electromagnetic forces. But no one knows which, if any, of these "grand unified theories," or GUTs, is most likely to be true. Furthermore, no attempts have been made to include the fourth force, gravity, within this theoretical framework. The reason for this is that in a certain sense gravity is more complicated than the other forces. According to Einstein's general theory of relativity, the gravitational field itself creates additional gravitational force. One could say that gravity gravitates. It is this complex behavior that has prevented scientists from finding any renormalizable theory of gravitation.

However, there is another approach that might eventually allow physicists to discover a theory of all the forces, and eliminate dependence upon the renormalization process as well. It is called superstring theory.

THE KALUZA-KLEIN THEORY

In 1919, the Polish physicist Theodor Kaluza created a version of Einstein's general theory of relativity that attempted to explain both gravitational and electromagnetic forces. He then sent it on to Einstein, hoping to get the latter's approval. Einstein liked some of Kaluza's ideas but suggested he do more work on the theory before it was published. Since in those days a paper could be submitted to a scientific journal only if a well-known scientist endorsed it, Kaluza's theory remained unpublished for two years.

Then, in 1921, the German mathematician Hermann Weyl propounded a theory in which he attempted to unify gravity and electromagnetism in a somewhat different way. Einstein was apparently not impressed by it. "Your approach seems in any case to have more to it than the one by H. Weyl," he wrote to Kaluza. Then he recommended Kaluza's paper for publication.

Kaluza's theory was a five-dimensional theory. It began with the assumption that there were four dimensions of space and one of time. Adding the extra dimension made it possible to view electromagnetism and gravity as two different aspects of the same basic force. The theory had a number of drawbacks. For example, it could not account for quantum phenomena, and it did not explain why the extra dimension of space was not observed.

Some of the theory's shortcomings were remedied by the Swedish physicist Oscar Klein in 1926. Klein showed that a quantum version of Kaluza's theory might be possible and suggested a reason why the extra spatial dimension—if it really existed—was not seen. It might be compacted, or "rolled up," to very small dimensions, something of the order of 10^{-32} centimeters,* which is about 10^{20} times less than the diameter of an atomic nucleus.

In order to see what Klein meant, imagine that you roll a two-dimensional sheet of paper into a tube. Imagine, next, that the tube can

---

* 10^{-32} is the number 1 divided by 10^{32}. It can be written as a decimal with 31 zeros following the decimal point before the numeral 1.

be rolled up as tightly as you want; its diameter can be made arbitrarily small. Under such circumstances, the tube will diminish in width until it begins to resemble a long, thin rod. One of the original two dimensions will have almost disappeared.

In Klein's theory the entire universe has a circumference of 10^{-32} centimeters along one of the four spatial dimensions, while the other three remain normal. If any object could travel along the compacted dimension, it could circumnavigate the universe while traveling a distance many times smaller than the diameter of an atomic nucleus. Naturally, the extra dimension, if it existed, could not be observed. It was not possible to see anything so small in Klein's time, and it is not possible today. At present, the most powerful accelerators can probe the nature of matter down only to a level of about 10^{-17} centimeters. The number 10^{-32} is 10^{15} (1 million billion) times smaller. An accelerator that could "see" objects of that size would have to be roughly a million billion times more powerful than any that exist today.

After Klein published his results, both Einstein and Pauli did further work on the theory, but their efforts were soon abandoned, and the Kaluza-Klein theory was forgotten. It remained forgotten for almost a half-century, until the idea of extra spatial dimensions was revived by scientists working on the theory of superstrings.

SUPERSTRINGS

If five space-time dimensions are needed to unify two of the forces, it seems reasonable that we might need even more if we want to account for all four of the forces of nature. In fact, contemporary superstring theories use quite a few more. They are generally formulated in ten dimensions— nine spatial dimensions and one of time. The extra six spatial dimensions are compacted to a size of about 10^{-33} centimeters.

In superstring theory, particles are not points. They are viewed as little vibrating loops, or superstrings, that have sizes that are roughly equal to those of the compacted dimensions, 10^{-33} centimeters. Since the particles are not points, there is hope that if a successful superstring theory is ever discovered, infinities will not appear and renormalization will not be necessary.

But that is a big *if.* Superstring theory has been described as twenty-first-century physics discovered by accident in the twentieth century. It is mathematically very complicated. Furthermore, many different super-string theories are possible. As I write this, there are five theories, and each theory has many thousands of different possible solutions. The number of theories is multiplied by the fact that where one extra dimension can be rolled up in only one way, six extra dimensions can roll up in and around one another in numerous ways. Thus the number of different possible versions of superstring theory is staggering.

Superstring theorists have made some progress. However, it is generally agreed that decades may pass before a theory that accurately describes the subatomic world is found, if indeed that ever happens. And even if such a theory is found, it will most likely never be possible to perform an experiment that confirms that superstrings really exist. It is not even certain that a particle accelerator the size of a galaxy would be capable of probing the structure of matter to such a level. On the other hand, a successful superstring theory would most likely solve the problem of the infinities that are encountered in quantum field theory while at the same time providing a unified explanation of the four forces.

A THEORY OF EVERYTHING
OR A BLIND ALLEY?

Among physicists there is no general agreement as to whether the search for a usable superstring theory is likely to succeed. Some liken the theoretical discovery of superstrings to the discovery of quantum mechanics in 1925. Others think that the superstring theorists have involved themselves in a futile pursuit. Feynman, for example, bluntly called super-string theory "nonsense," and Stephen Hawking has added his voice to those of the skeptics. Nobel laureates are found on both sides of the controversy. Some criticize the theory, while others call it "the only game in town" and express the hope that a theory will eventually be found that explains "everything."

When physicists speak of a theory of "everything," they mean a theory that will explain the four forces. For if the forces of nature are understood, then all other laws of physics could presumably be derived from the theory. Naturally, complex superstring equations would not be used for everyday calculations, such as finding the pressure in a helium balloon or the heating of the element of a lightbulb, but it could at least be done in principle.

On the other hand, many physicists feel that no theory of everything will ever be found. They feel that nature is inexhaustible, that physicists will find theories that are better and better approximations, just as they have in the past, and that this process will never come to an end. One might even say that they believe the number of different possible levels of understanding that can theoretically be attained is infinite.

A LAST-MINUTE NEWS BRIEF

After I wrote this chapter, I learned that experiments performed at the Fermi National Accelerator Laboratory (commonly called Fermilab) in Illinois suggest that quarks might have smaller constituents. A large team of experimenters caused protons and antiprotons to collide with one another at high energies. In such collisions, the quarks that are the constituents of the protons and antiprotons interact with one another. Thus forces between quarks can be studied.

The idea behind the experiment was the same as that underlying Rutherford's experiments in the early twentieth century. The major difference was that the protons and antiprotons collided with energies more than 100,000 times greater than that of Rutherford's alpha particles. Therefore the nature of matter could be probed at a much smaller level.

The experiments found that when the structure of matter was probed at dimensions ranging from about 100 to about 1,000 times smaller than a proton, the results agreed with existing theory. At dimensions smaller than this, however, the data began to diverge from theoretical predictions. One possible interpretation of this result is that quarks are made up of smaller particles, and that these subquark particles were interacting with one another.

As of this writing, it is too early to know whether this idea will be confirmed in further experiments, or what the implications of this discovery might be. It will likely be a long time before any definite conclusions are reached. Years of effort are required to extract the relevant information from the massive quantities of data produced in modern particle accelerator experiments. For example, the data suggesting that quarks might have components was obtained in experiments performed in 1992 and 1993, but the results were not announced until early 1996.

At present, there are only hints that quarks have components, and it is not really known what the implications of this might be. So let's look at some possibilities:

1. It is possible that this idea will not be confirmed. In that case, nothing will have changed. A number of cases in which experimental physicists thought they had discovered some new particle later turned out to have been mistakes. And in this case, no new particles have as yet been found. There are only hints that they might exist.

2. If quarks do turn out to have constituents, it opens up the possibility that these constituent particles might not be elementary, either—they may have their own constituents. In such a case, it is conceivable that the number of levels might be infinite. Every particle could be made up of particles that are still smaller. Atoms are made of electrons, protons, and neutrons. Neutrons and protons are made of quarks. Quarks are made of some kind of new particle. And there is no reason to stop there. Physical reality could be like an onion, with an infinite number of layers.

But if quarks have constituents, we will most likely never know whether or not there is some fundamental level at which particles are truly elementary. The more deep the probe into the structure of matter, the greater the quantities of energy required. There are certain practical limits, which are being approached right now.

3. Two basic families of particles have been discovered, the quarks and a group of particles called leptons. Each family has six members. The electron is a member of the lepton group. If the quarks turn out to have constituents, then it is reasonable to suspect that the leptons, including the electron, might not be elementary, either. If the electron has constituents, then, like the proton and the neutron, it has some finite size;

its constituent particles would interact with one another in some small but finite volume of space. Gaining an understanding of the lepton's constituents might point the way to the discovery of methods that would eliminate the infinities in QED.

4. It is not clear what the implications of all this are for superstring theory. It may turn out that the superstring theory will be discredited. Or it may be that the theory will find itself able to explain the behavior of the constituents of quarks, if such things exist. One thing is clear—these hypothetical quark constituents are almost certainly not superstrings, which would be too small to be seen.

7

THERE WAS A YOUNG LADY
NAMED BRIGHT

> There was a young lady named Bright
> Who traveled much faster than light.
> She started one day
> In the relative way,
> And returned on the previous night.

THIS OFTEN-QUOTED LIMERICK, originally published long ago in the British magazine *Punch* when Einstein's theories were first becoming known to the general public, gives a pretty accurate description of one of the implications of Einstein's special theory of relativity. The theory tells us that if anything—either a material object or a signal—could travel faster than the speed of light, then it would also be able to travel from the future into the past.

Like many surprising conclusions in physics, the idea that faster-than-light travel can, under certain circumstances, also be time travel can be deduced from some very simple assumptions. The special theory depends upon just two. The first is that the velocity of light, as measured by any observer, is always the same. The second is that the laws of physics will appear to be the same to any observer in a state of uniform motion. Here, "uniform" means "at a constant speed in a fixed direction." The distinction between uniform and nonuniform motion is significant. For example, a passenger in an airplane that is moving at a constant velocity in a straight line feels the same force of gravity that she feels on the surface of the earth, and she can walk up and down the aisle as though she were walking down the aisle of an auditorium. But if the plane suddenly descends when it encounters turbulence, this can change. Under extreme conditions, a plate of food can even seem to go flying upward.

The idea of relativity is actually nothing very new. Galileo and Newton knew that motion was relative. They were aware that a passenger on a moving ship sailing across a calm sea can, if he wants, consider the ship to be at rest. As Galileo pointed out, an object dropped from a mast will seem to drop directly downward toward the deck whether the ship is moving across the surface of the ocean or not. The only thing that matters is the motion of the object relative to the ship.

Indeed, we are all natural relativists. Someone who is sitting in a chair will generally consider himself to be "motionless," even though the earth is rotating on its axis and revolving around the sun, while the sun revolves around the center of our Milky Way galaxy, which is moving with respect to other galaxies in space. None of these motions is uniform. Circular motion, for example, is not uniform because it is not motion in a straight line. However, these motions come close enough to uniformity for the purposes of everyday life.

A PROBLEM WITH ELECTRODYNAMICS

When Einstein began to think about relativity during the early years of this century, there was one glaring exception to the principle that the laws of physics should always be the same for any state of uniform motion. The laws of electricity and magnetism, or electrodynamics, were not always the same. For example, if a magnet is moved, an electric field will be created. This electric field can, in turn, induce an electric current in a nearby wire. This is the principle on which the generation of electricity is based. Electrical generators contain rapidly moving magnets.

Similarly, if the magnet is motionless but the wire is moved, an electrical current will also be set up. However, in this case, the laws of electrodynamics described the phenomenon differently. They still predicted the appearance of a current, but they seemed to say that there was no electric field; there couldn't be if the magnet didn't move.

In Einstein's day, everyone—or at least all physicists—knew there was something funny about this. If relative motion was all that mattered, why should physical laws describe the two cases differently? Everyone knew this, but no one commented on it; the contradiction was conveniently ignored. As Ronald W. Clark, a biographer of Einstein, put it, to raise questions about this difficulty "was to spit in a sacred place."

But Einstein was never one to blindly accept authority, and he did raise awkward questions. While he was still an adolescent, he noticed there was something paradoxical about the accepted electromagnetic theory. It was well established that light waves were made up of rapidly oscillating electric and magnetic fields. But suppose one followed a light wave at the speed of light? In that case, the fields would seem to be motionless, just as an ocean wave would seem to be motionless if one flew above it at the same speed that it was traveling. But motionless electric and magnetic fields did not exist in nature in the absence of electric charges and magnets. Thus, if travel at the speed of light was possible, something that was clearly impossible would be seen.

In 1905, Einstein published a paper with the modest title "On the Electrodynamics of Moving Bodies" in which he showed that the contradictions in electromagnetic theory could be removed if one made the assumption that the velocity of light would seem to be the same to any observer in a state of uniform motion. This paper laid the foundations of Einstein's special theory of relativity. It contained all the important results of the theory except the one concerning the equivalence of mass and energy. Einstein's famous equation $E = mc^2$ did appear in a second paper on relativity, published later that year.

The idea that the speed of light should always be the same seems a little startling at first. For example, suppose a spaceship is motionless with respect to a star. If the people on the spaceship measure the velocity of the light that reaches them, they will find that it is traveling at 300,000 kilometers (186,000 miles) per second. If the spaceship is then accelerated to a velocity of half the speed of light toward the star, or away from the star, they will still obtain the same result. Similarly, if a beam of light is emitted by the spaceship, it will move away at the same velocity of 300,000 kilometers per second whether the ship has been accelerated to some high velocity or not.

That such things should be the case seem an affront to common sense. However, as we saw in chapter 6 on quantum mechanics, counterintuitive phenomena are commonplace in a realm so far removed from the world of everyday experience. Quantum mechanics tells us that we cannot expect objects in the subatomic world to behave the way that large macroscopic objects behave. Similarly, relativity tells us that objects that move at velocities approaching that of light will not act in the same manner as those that move relatively slowly. Of course, this also applies to light itself, which moves at light velocity by definition.

THE MICHELSON-MORLEY EXPERIMENT

Relativity was not a speculative theory. As Einstein liked to point out, it was designed to explain observed experimental facts. When Einstein propounded the theory, evidence already existed that seemed to indicate that state of motion did not affect measurements of the speed of light. In 1887, the German-American physicist Albert Michelson and the American chemist Edward Morley performed a much-discussed experiment in which they attempted to compare the velocity of light in different directions with respect to the motion of the earth. Direct measurements of the velocity of light in different directions would have been of little use. These could not be performed with sufficient accuracy to determine if there was any discrepancy. Hence Michelson and Morley used sets of mirrors that caused a beam of light to travel in two different directions at the same time. Their experiment was designed to detect differences in light velocities, which was something that could be measured. The following analogy should clarify what they were doing. Suppose that you are watching some runners in a 100-meter dash. If you don't have a stopwatch, it would be difficult to measure the average speeds at which they run. On the other hand, it's easy to determine if one runner is a little faster or a little slower than another. You only have to see who crosses the finish line first.

Michelson and Morley fully expected to obtain a positive result. At the time, physicists believed that light was carried by a substance called the ether, which filled all space. They reasoned that if light consisted of electromagnetic vibrations, then there naturally had to be something for it to vibrate in. Every other kind of oscillation they knew of required movement of some physical substance. Sound waves, for example, consisted of movement of air molecules. There was no reason to expect light to be any different.

To their surprise, Michelson and Morley obtained a negative result. They failed to find detectable differences in the speed of light as it traveled in different directions. It didn't make any difference whether a light beam traveled in the direction that the earth was moving, or in the opposite direction, or in a perpendicular direction. Its velocity was always the same. And since the light beam was presumably carried by a stationary ether through which the earth was traveling, this was a puzzle indeed.

The ether theory was not to survive for long. As Einstein himself pointed out in his paper on relativity, the failure of attempts to detect the ether rendered it an unnecessary hypothesis. In fact, the idea of an ether would not have been consistent with relativity. If such a thing existed, motion would not be relative. Any observer would always be motionless with respect to the ether or be moving through it. Einstein solved the problem of the ether simply by dismissing the idea. Of course he was right. Today, physicists see no contradiction in the idea that electromagnetic waves can travel through empty space.

CHANGING PERSPECTIVES IN TIME

The idea that the velocity of light is the same for all observers leads to some surprising results. For example, two events that one observer finds to be simultaneous will, in general, not appear to happen at the same time from the point of view of another observer. If one observer finds that

THERE WAS A YOUNG LADY NAMED BRIGHT • 125

event A and event B occur simultaneously, a second will find that A happens first, while a third will conclude that B was the first.

This can be illustrated by a thought experiment. Suppose that a ship is sailing past a coast on a dark night, and that lightning strikes in two places along the shore. Suppose, next, that an observer is standing on the shore at a point midway between the two lightning flashes. Since the light from each of the flashes reaches him at the same time, he concludes they are simultaneous.

Now suppose that the ship is also equidistant from the two flashes, and that it is moving away from one lightning strike in the direction of the other. Since the ship is moving away from one set of light waves and toward the other, an observer on board will not see the two flashes simultaneously. One will arrive a tiny fraction of a second sooner. But the observer on the ship has just as much right to consider himself to be motionless as the person on the shore. And if one flash is seen before the other, he is perfectly justified in concluding that this lightning strike happened first.

In practice, the difference between the times of arrival of the two flashes would be too small to measure. However, if two flashes of light were viewed by an observer on a spaceship traveling at a considerable fraction of the speed of light (compared with some "stationary" observer), the difference could be quite large. If the flashes were far enough apart, and his velocity great enough, the difference could be a matter of years. Yet the "stationary" observer could see them as happening at the same time.*

When a three-dimensional object is viewed from different angles, the visual images that it produces change. We sometimes say that we can view it from different perspectives. Einstein's theory of relativity tells us that we can also see time from different perspectives. The temporal order of two events may seem different to different observers according to their state of motion.

But the temporal order of two events cannot always be reversed. If they

---

* Here, I put the word *stationary* in quotation marks because it is an arbitrary concept. Any observer who is not being accelerated can consider himself to be stationary.

are close enough to one another in space, or distant enough from one another in time, all observers will see one happen before the other. For example, there is no possible state of motion that would cause an observer to conclude that an atomic bomb was dropped on Hiroshima before the attack on Pearl Harbor. And there is no possible state of motion that will cause an observer to see a batter hit a baseball before the pitcher makes his delivery. Or at least this cannot happen if the observer is moving at a velocity less than the speed of light.

THE FASTER-THAN-LIGHT FASTBALL

Let's return to our Celestial League. The Babe is feeling frustrated. It is the sixth inning of the second game of a doubleheader, and he hasn't had a hit all day. The pitcher he is facing is one of the best. When this pitcher is really hot, no one can hit him, for his fastballs travel faster than the speed of light.

The problem with faster-than-light fastballs is that the batter never sees them until they are in the catcher's mitt, for they outrun the light that is reflected from their surfaces. In fact, a faster-than-light fastball appears to go backward from the catcher to the pitcher's hand; the light that was reflected from the ball's surface when it was halfway to the plate arrives *after* the ball itself, and the light that was reflected at the moment it was thrown arrives later still.

From the standpoint of the ball, things also look odd. As the ball makes its way toward the batter, it outruns the light reflected from the pitcher. If there were an observer on the ball (we can assume, if we wish, that a small angel has decided to hitch a ride), he would first see the ball released from the pitcher's hand, and then, as light from later moments arrived, he would see the pitcher go through his windup backward.

Well, as everyone knows, God is a great Yankee fan. When He sees that His favorite player has not gotten a hit, He decides to guide the Babe's bat. The next time a faster-than-light fastball is thrown, the Babe finds himself meeting it with a faster-than-light swing. But instead of hitting a home run, the Babe hits one of those towering pop flies he is known for.

The ball flies upward, still traveling faster than light. As the ball progressively catches up with light rays that left the playing field at earlier moments in time, the angel who is riding the ball sees the entire game being played in reverse as she looks down. Fly balls pop out of outfielders' gloves, arc toward bats that are swung in reverse, and then bounce into the pitchers' hands. Meanwhile runs gradually disappear from the scoreboard.

And of course the angel never sees the umpire, who is standing there scratching his head, wondering whether a ball that left the stadium in an upward direction should be called a home run or an automatic out.

INFINITE ENERGY

The effects of traveling faster than light, as the intrepid Ms. Bright discovered, are more than just illusion; if such a thing were possible, we really could travel into the past. And of course, if this were possible, serious paradoxes would result. For example, suppose it were possible to send a robot on a faster-than-light journey. The robot could be programmed to leave on Tuesday and arrive the previous night. The robot could then dismantle itself—or better yet, commit suicide by sending surges of high-voltage electricity through its delicate positronic brain—a day before it left, making the journey impossible. If you can travel backward in time, you don't have to kill your grandmother before your mother was born to eliminate yourself. You can simply do yourself in.

Of course, traveling in time, if it were possible, would never become a popular method of suicide. At best, an infinite regress would result. If you go back in time to kill yourself, then you would not be alive to make the journey. But if you were not alive on Tuesday, you could not go back and kill yourself on Monday. Thus you would survive until Tuesday, and you could go back to kill yourself. But if you did go back and kill yourself after all. . . . The sequence is unending; it goes on and on and on.

Naturally, if you were able to travel faster than light, it would be possible to go to the future as well as to the past. But nobody talks much about travel to the future, or at least physicists don't. Paradoxes are not associated with it. Traveling to the past would allow us to alter the present. Traveling to the future would not.

In order to see why relativity forbids faster-than-light travel, I will examine the situation from two different viewpoints. After all, according to the theory of relativity, the outlooks of all observers are equally valid.

If you were traveling in a spaceship that has accelerated to a high velocity, you would never catch up with a ray of light. No matter how fast you go, that light ray would still appear to be racing ahead of the ship at a velocity of 300,000 kilometers per second.

Let's also look at the situation from the point of view of an observer on earth. According to the theory of relativity, when an object is accelerated to a high velocity, it appears as though the object's mass has increased. For example, a 100-kilogram object that is traveling at 90 percent of the speed of light will appear to weigh just a little less than 230 kilograms. If the velocity is 99 percent of that of light, its mass will increase to a little more than 700 kilograms. Naturally, an observer on a spaceship would not notice any such effect. In relativity, many things look different when observed from different points of view. The relativistic mass increase, by the way, is an effect that has been observed experimentally. Subatomic particles that are accelerated to velocities approaching the speed of light in modern particle accelerators exhibit precisely the kind of mass increases that relativity predicts.

From the viewpoint of an earthbound observer, as the spaceship goes faster and faster, it becomes heavier and heavier. This makes it progressively more difficult to add each additional increment of velocity because more energy is required. A heavy object, after all, has more inertia than a relatively light one. It is a lot harder to push an automobile than it is to push a bicycle. If the mathematics are worked out in detail,* one finds that an infinite amount of energy is required to accelerate an object to light velocity. Since infinite quantities of energy are not likely to be available anytime soon, faster-than-light travel isn't likely, either.

---

* Since the formula really isn't very complicated, I'll give it here for the benefit of the mathematically inclined. The mass of a moving object is equal to its rest mass—the mass when it is not moving—divided by the square root of the quantity $1 - v^2/c^2$. Here, v is the velocity of the object.

It appears that the appearance of infinite quantities in the equations of physics does not always lead to catastrophe. The conclusion that the speed of light is a limiting velocity is an example. Here, the appearance of infinity tells us that something cannot happen, and the conclusion that something is impossible can be as important as the discovery that something is possible. Another example, one that we have already encountered, is the Heisenberg uncertainty principle, which states that it is impossible to determine exactly the position and the momentum, or the energy and time of existence of a particle at the same time. As we saw, this led to the important conclusion that virtual particles had to exist.

$$E = mc^2$$

$\mathbf{T}$he phenomenon of relativistic mass increase and the consequent existence of an "infinity barrier" has far greater importance than it appears to have at first. It was the former that led Einstein to the famous formula $E = mc^2$. If expenditure of energy leads to increased mass, it seems to follow that the two must somehow be equivalent. This prompted Einstein to make the assumption that the energy of a body was *always* equal to mc^2, whether it was moving or not. He found that this led directly to the correct formula for relativistic mass increase. The natural conclusion is that if infinity did not prevent objects from traveling as fast as light, then E would *not* equal mc^2, and we would live in an entirely different kind of universe. The fact that infinite quantities of energy cannot exist has very real implications for our understanding of the natural world.

A CLASSICAL THEORY

$\mathbf{T}$he discovery of the theory of relativity was surely one of the greatest achievements of twentieth-century physics, but it was not as revolutionary as Bohr's assumption that orbital electrons did not radiate energy. Bohr's theory led to the discovery of the strange world of quantum mechanics, while Einstein's is really an extension of classical nineteenth-century physics. Though both theories changed the way physicists viewed the natural world, Einstein's elaborated on existing ideas more than it changed them. Interestingly, the conservative Planck, who was horrified by his own quantum hypothesis and labored for years trying to find a way around it, enthusiastically advocated relativity theory.

It has been said that if Einstein had not discovered the theory of relativity, someone else would have done so within a few years. The French mathematician Henri Poincaré discovered many of the ideas associated with relativity (but not $E = mc^2$) independently of Einstein. The Dutch physicist Hendrik Lorentz was so important a precursor that references were sometimes made to the "Einstein-Lorentz" relativity theory in the years after Einstein's theory was established. Both Poincaré and Lorentz were contemporaries of Einstein, though a little older in age. Until Einstein came along, Lorentz was considered to be the most eminent theoretical physicist of the day. His colleagues sometimes referred to him as "the great Lorentz."

To point out these anticipations is not to denigrate Einstein's achievement. He was by far the greatest physicist of his time. He made not just one but many contributions to physics that were important enough to have won him the Nobel Prize.* And, although the special theory probably would have been discovered by someone else if Einstein had never been born, it is doubtful that his theory of gravitation, the general theory of relativity, which he published in 1915, could have been developed by anyone else then alive.

Theoretical attempts to circumvent the light barrier generally make use of Einstein's general theory. Physicists have sometimes wondered if this second relativity theory might allow faster-than-light travel and time travel, even though the special theory did not. Somehow this seems appropriate. It seems that Einstein's stature as a physicist was so great that Einstein must be made use of in order to have any hope of refuting Einstein.

---

* When Einstein received the prize in 1922, it was awarded for his contributions to quantum theory, not for his work on relativity.

THE GERMAN LANGUAGE
AND "CURVED" SPACE

Einstein's declaration that no one could travel faster than light never deterred science fiction writers from introducing space travel into their stories. I read sci-fi voraciously when I was an adolescent. Even then, the idea of a "space warp" was commonly used to circumvent the light barrier. Something similar is seen today in the various *Star Trek* series, where ships are equipped with "warp engines" that enable them to travel around the universe at "warp speed."

The idea of a "warp" is related to the idea of "curved" space mentioned so often whenever Einstein's general theory of relativity is discussed. (I will be discussing this theory in detail in the next two chapters.) The original idea seemed to be that if space could be curved, then there was probably some way to warp it a lot in order to get somewhere fast.

Space is not a "thing" that can be bent or curved. When scientists speak of "curved" space, they mean that gravitating bodies change the geometry of space so that it is not quite like the Euclidian geometry we are taught in high school. In Euclidian geometry, the angles of a triangle always add up to 180°. In non-Euclidian geometry, the sum of the angles can be either more or less. The surface of the earth provides a good example of this. The lines of latitude are all perpendicular to the equator; that is, they intersect the equator at angles of 90°. These lines all come together at the North and South Poles. Thus two lines of latitude and a section of the equator form a triangle. Since the two 90° angles at the equator add up to 180° and the angle at the Pole has some magnitude greater than zero, the sum of the angles is more than 180°.

The surface of the earth is curved in a third dimension of space. On the other hand, there is no extra spatial dimension in which three-dimensional space can be curved, so the situation is somewhat different. It would be most accurate simply to speak of Einstein's space as having a non-Euclidian geometry. However, it so happens that most of the original papers on non-Euclidian geometry were written in German, and the

German language does not permit the formation of an adjective corresponding to the English word "non-Euclidean." Consequently, the German mathematicians who wrote on the subject got into the habit of using the term curved instead. Since Einstein's papers on relativity were also written in German, the phrase was carried over into English when they were translated.

The term "curved space" is used so often that I would not want to argue for its elimination. The usage in fact is perfectly acceptable as long as we remember that we are resorting to analogy, and that space is not curved in the same way that a material object might be. I will be using the term myself in subsequent chapters of this book. On the other hand, the term space warp, as it is used in science fiction does not correspond to anything real. If the German language were a little different—if it did possess a term corresponding to "non-Euclidian"—then science fiction writers most likely would have had to invent some other way to get around the universe.

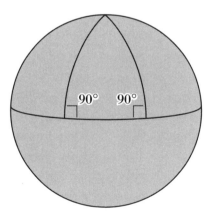

Figure 7. The geometry of the surface of the earth is non-Euclidian. The sum of the angles of a triangle is always greater than 180°. For example, the sum of the angles of a triangle formed from two lines of latitude—which meet at the North Pole—and a section of the equator must always be more than 180°, because the lines of latitude cross the equator at 90° angles.

YOU CAN'T GET THERE
FROM HERE, OR CAN YOU?

General relativity tells us that we live in a world with a funny kind of geometry that can cause a traveler to wind up in places we would not expect him to be going—and this might just include times in the past or in the future. After all, like the special theory, general relativity is a theory of space *and* time. It tells us that gravitational effects can cause time to be seen from a different perspective than the ordinary one.

We are normally not aware of relativistic effects. Only scientists observe particles traveling at velocities approaching that of light. Similarly, the space and time distortions predicted by general relativity only become significant in strong gravitational fields. Though experimental tests of general relativity have been conducted on the surface of the earth, the effects that were measured were too small to be seen by anyone but physicists who have constructed very precise measuring apparatus.

So don't expect to find yourself traveling in time anytime soon. On the other hand, general relativity has led some scientists to speculate about the question of whether a time machine, or a mechanism for traveling at velocities greater than light, could somehow be constructed.

The California Institute of Technology physicists Kip Thorne, Michael Morris, and Ulvi Yurtsever have suggested that a highly advanced technological civilization might indeed have the capability of constructing such a time machine. Yes, this sounds like science fiction, and in a sense it is. When astronomer Carl Sagan was writing his novel *Contact*, he asked Thorne to suggest a plausible method for interstellar travel. The idea aroused Thorne's interest, so together with Morris and Yurtsever, two of his Ph.D. students, he began to investigate the properties of theoretical objects called wormholes.

Wormholes are "bridges" that would connect widely separated regions of space. Anything that traveled through a wormhole could emerge in a region of space millions of light years away. Passing through a wormhole would not take very long. So, in effect, that object would have traveled to its destination at speeds far greater than that of light. Although wormholes have never been observed, solutions to the equations of general relativity allow for their existence. It is not very likely, however, that we will ever encounter macroscopic wormholes like the one encountered in the television series *Star Trek: Deep Space Nine*. Calculations have shown that if such wormholes were to exist today, the universe would have to have had a very unlikely, bizarre character when it was created in the big bang.

On the other hand, it is possible that microscopic wormholes with dimensions of the order of 10^{-33} centimeters might well exist. If so, they would rapidly pop into existence in all regions of space, and then disappear again within a very short time, in a manner similar to that of virtual particles. Now, 10^{-33} centimeters is many orders of magnitude smaller than the dimensions of an atomic nucleus, which are about 10^{-12} centimeters. There is no hope of observing such tiny objects with present-day technology. In fact, it isn't clear that they could ever be observed.

This didn't deter Thorne, Morris, and Yurtsever, however. The question they were asking was this: If some advanced civilization was able to find these microscopic wormholes, capture them before they disappeared, and enlarge them to macroscopic dimensions, could they then be used for interstellar travel? They found that this was indeed theoretically possible, though the difficulties would be formidable. Looking at the simplest case, that of a wormhole with a circular cross section, they calculated that some "exotic material" or "exotic field" capable of withstanding pressures of approximately 10^{37} pounds per square inch would be required if the wormhole was to be kept open. It could turn out to be theoretically impossible to create such a material or field, the authors admitted. However, the creation of traversable wormholes had to be considered at least a possibility.

The idea intrigued other theoretical physicists, and numerous papers were written on the subject. A great number of questions needed to be considered. For example, would it be more feasible to construct a wormhole with some other shape? Would this make the pressure smaller? Would travelers who attempted to journey through a wormhole be subjected to forces or radiation that might prove fatal? Would the gravitational influence of a spaceship through a wormhole render it unstable and cause it to close before a ship could get through? How long would a journey through a wormhole take? How great would the accelerations be for space travelers?

Obviously, it is not clear if wormhole travel will ever become a real possibility. The idea does conjure up the image of a kind of cosmic railway station, where instead of trains, wormholes take travelers to distant parts of the universe. If it turns out that such a project cannot be carried out in reality, it still provides a useful theme for writers of science fiction.

BACK TO THE PAST AGAIN

If macroscopic wormholes could be constructed and maintained, they could be used as time machines. Theoretical calculations indicate that if a wormhole were constructed in such a way that its two mouths (think of a wormhole as a kind of tunnel; like a tunnel, it has mouths) were relatively close to one another in space, and if one mouth was moving rapidly while the other was static, then a traveler who passed through the wormhole would make a journey in time rather than in space. Of course, this creates all the usual paradoxes. For example, presumably a billiard ball could enter a wormhole in such a way that when it emerged in the past, it would strike itself and thus prevent itself from entering the wormhole in the first place. But if the billiard ball never entered the wormhole, it couldn't go into the past to prevent itself from making the journey.

Speculation of this sort has something of the nature of a game. It is, however, a game with a serious purpose. Real advances in scientific knowledge come about when the laws of physics are pushed to their limit. It may be unlikely that traversable wormholes will ever exist, but, if physicists discover they are possible in principle or, alternatively, that the laws of physics forbid their existence, something will have been gained.

At times physicists' theoretical game playing seems to elicit something resembling real alarm within the community of theoretical physicists. When the Princeton physicist J. Richard Gott proposed a somewhat more plausible method for traveling in time, there was an immediate reaction. Certain other theoretical physicists attempted to show that his idea would not work, and a brief but intense controversy resulted.

Gott's idea made use of cosmic strings, strange objects that might have been created early in the history of the universe. Cosmic strings have never been observed, but they are a theoretical possibility. Cosmic strings are not related to superstrings. If they exist, they are long, filamentlike concentrations of energy. It has been calculated that a piece of cosmic string the size of an atom would weigh about a billion tons, and that a segment as long as a football field would weigh as much as the earth.

Such objects would not only exert strong gravitational forces. According to the general theory of relativity, they would also strongly distort time perspectives in their vicinity. Furthermore, it has been shown that if cosmic strings exist, they must move about the universe at velocities approaching that of light. They are relativistic objects in every sense of the term.

In 1991, Gott published a paper in which he claimed to have shown that if two cosmic strings hurtled past each other in opposite directions, and if a spaceship followed a certain trajectory in the vicinity of the strings, then it could travel into the past without exceeding the velocity of light. Furthermore, it could follow a path that would bring it back to its starting point before it had started its journey.

Gott's paper was followed by responses from a number of different physicists. Some of them presented arguments purporting to show that Gott's method would not work. Then the American physicists Stanley Deser and Roman Jackiw and the Dutch physicist Gerard 't Hooft claimed they had found a fallacy in Gott's argument. Naturally, Gott disagreed, contending that it was his theoretical opponents who had made a mistake. Gerard 't Hooft responded by writing another paper in which he claimed to have demolished Gott's theory.

Again, Gott accused his opponents of using fallacious arguments, and cited a paper written by the California Institute of Technology physicist Curt Cutler. Cutler's paper, Gott said, showed that time travel was a real possibility. At this point, the battle was joined by Stephen Hawking, who joked that if time travel were a possibility, we would have encountered hordes of tourists from the future. Hawking then performed some calculations that showed, he said, that there would be energy buildups that would destroy Gott's time-travel loopholes as soon as they were created. At this point, the controversy seems to have died down, though there were still uncertainties as to whether Gott's scheme would or would not work. However, physicists have not stopped looking for solutions to the equations of general relativity that would allow time travel. Though few believe that travel in time is a real possibility, they continue to plumb the depths of Einstein's theory.

TACHYONS

The possible existence of faster-than-light particles, or tachyons, was suggested long before the hypothetical time travel schemes I described above were developed. I saved a discussion of them for last because tachyons would not permit actual time travel; at best they would make it possible to send messages into the past.

I previously pointed out that nothing can travel faster than light. As it turns out, this might not be entirely true. During the mid-1960s, physicists Gerald Feinberg and George Sudarshan showed independently that special relativity does not rule out the existence of faster-than-light particles; it only implies that if such particles did exist, they could never travel at velocities *less* than that of light. They would meet the infinity barrier from the other side.

If tachyons exist, they must behave in some very odd ways. For example, if a tachyon lost energy, it would move faster, not more slowly. And if the energy of such a particle were to fall to zero, it would travel through space at infinite velocity. But of course this "oddness" is no argument against tachyons' existence. After all, the idea certainly sounds less strange than Dirac's idea of a sea of negative electrons, or that of an electron possessing infinite charge and infinite mass.

If tachyons exist, and if beams of tachyons could be created in the laboratory, then it would presumably be possible to use tachyons to send messages into the past. Not surprisingly, this could lead to paradoxes as

severe as those associated with time travel. For example, it would be possible to send the design for some novel kind of technology back to a time
several years or even decades earlier. The technology could then be "invented" in the past, and the reason that we would have it today would be
that it was developed back then. But in such a case, where would the
original idea have come from? Similarly, we could transmit Shakespeare's plays back to Shakespeare at times before he wrote them. If
Shakespeare then copied them down, would this mean that the plays
came into existence out of nothing? Under such circumstances, the question of the authorship of Shakespeare's plays really would be a problem.
Of course, if we pursue this scenario, it is necessary to think of some way
to get the messages to Shakespeare, who is not likely to possess a tachyon
receiver.

Other possible paradoxes resemble those associated with time travel.
Suppose a friend dies in an automobile crash, and I send a message to the
past warning him not to drive anywhere on the fatal day. If he takes my
advice, he will not die in a crash, and there will be no reason for me to
send the message in the first place.

After Feinberg and Sudarshan showed that the existence of tachyons
might be possible, a number of experiments were performed to detect
them. All produced negative results, and even Feinberg no longer believes that tachyons are very likely to be real. However, I should point out
that the possibility of their existence has not been definitely ruled out. It
is even possible that they exist in great numbers, but they do not interact
with slower-than-light matter. In this case, it could not be said that they
were a part of *our* universe.

The question of the existence of tachyons has an interesting sidelight.
It raises the question: If we find that it is possible for something to exist,
should we assume that it does exist? California Institute of Technology
physicist Murray Gell-Mann has expressed this idea in a slightly different form, which is known as the Totalitarian Rule of Physics: "Anything
that isn't forbidden is compulsory." The Totalitarian Rule can be a useful
maxim for scientists working in the field of high-energy particle physics.
If a certain kind of interaction between particles is not forbidden by the
known laws of physics, one can reasonably expect that it happens. But is

the Totalitarian Rule always true? Might nature sometimes follow what one might call the Democratic Rule of Physics? This could be expressed: "Anything that isn't forbidden is permitted, but not required." So it is easy to see, if the Totalitarian Rule is always true, tachyons must be out there. If the Democratic Rule is true, this is not necessary.

WELL, IS TIME TRAVEL POSSIBLE, OR ISN'T IT?

As much as they may play around with the idea, most theoretical physicists hope that time travel isn't possible. If an individual or an object could travel into the past, this would upset accepted ideas about causality, and invalidate the laws of physics, which depend upon it. The only way to avoid this would be if some law of nature prevented anything that traveled into the past from altering the past, or a law that ensured that anything a traveler did would produce precisely the world from which he had come. But it is hard to see how a law of nature could prevent someone from killing his grandmother—or himself—if that is what he really wanted to do.

Stephen Hawking has another solution to the problem. He suggests that a "Chronology Protection Agency" prevents time travel. No, he is not suggesting that Time Police exist who prevent people from making journeys into the past. Hawking just enjoys expressing serious ideas in flippant ways. He only means that he suspects that the laws of physics operate in such a way as to render time travel impossible. However, it isn't very clear how his "chronology protection" would operate. It may be that all time travel mechanisms have features that would render them unworkable in practice, but it isn't easy to imagine what kind of natural law would cause such a situation to come about.

In any case, even though time travel cannot be absolutely ruled out, it certainly appears to be an unlikely possibility. In spite of all the attempts that have been made to surmount it, the "infinity barrier" erected by the special theory of relativity has proved to be a hard one to overcome.

8

SINGULARITIES

ABOUT FIVE BILLION YEARS from now, the nuclear fuel that keeps our sun bright and hot will begin to become exhausted. As it does, it will begin to grow rapidly in size. You might think that a dying star would become smaller and dimmer, but, at first, just the opposite happens. As stars grow older, the nuclear reactions they use to produce energy proceed at a more rapid rate. This causes their internal temperatures to rise, and the heat produces outward pressure that causes the stars to expand. Although our sun is only middle-aged, it already shows signs of aging. It is about two and a half times as bright as it was when the earth was formed about 4.5 billion years ago, and its diameter is a few percent larger.

As our sun enters its death throes, the changes it experiences will be even more dramatic. It will evolve into a hot, luminous red giant. It will swallow up Earth, and then expand still farther until its surface lies at about the present orbit of Mars. Only when the sun's nuclear fuel is completely exhausted will it begin to shrink. Once this process begins, it will continue until the sun becomes a tiny, dim white dwarf.

White dwarfs are dead stars. They glow only because they possess a great deal of residual heat. As this heat escapes into space over a period of billions of years, they grow progressively dimmer. Every white dwarf will eventually evolve into a black dwarf, a cold relic of a dead sun.

ELECTRONS MAKE A TABLE SOLID

To understand the properties of white dwarfs, it is necessary to know something about the role that electrons play in the structure of matter. It is the interactions between electrons that give matter its character. A table is solid because of the way its constituent electrons interact with one another. The nuclei of the atoms that make up the table never get close enough to one another to produce forces of any significance. It is the electrons that bind atoms to one another. The properties of liquids can also be explained by electron interactions; in this case, the electrons do not form permanent bonds; on the contrary, they slide past one another. Thus it seems reasonable that a white dwarf would have the properties it has because of the way its electrons interact.

When a star the size of the sun contracts into a white dwarf, the gravitational forces become very strong. The electrons are packed as closely together as they can be. As a result, the density of white dwarf matter is very great—a tablespoon of this material weighs several tons. A white dwarf can be thought of as a body in which an electron "gas" has been compressed by gravity. Again, atomic nuclei play no significant role; they are still relatively far apart.

A white dwarf cannot be an arbitrary size. In particular, it cannot have a mass that is greater than 1.4 times the mass of our sun. If it does, gravitational forces will become so strong that the pressure of the electron "gas" will not be able to prevent further collapse. The electrons and protons in the star will be "squeezed" together into neutrons, and a neutron star will be created.

Neutrons are generally stable when they are inside atomic nuclei. But some radioactive atoms emit beta particles, or electrons. When this happens, one of the constituent neutrons of a nucleus is converted into a proton. Free neutrons exhibit the same kind of behavior. On average, a free neutron will decay into a proton and an electron in about eleven minutes.

If a nuclear reaction is possible, then the reverse reaction is always possible too. In particular, if an electron strikes a proton with sufficient energy, the two can form a neutron. This is exactly what happens in a massive collapsing star. If the star is big enough, there will be enough gravitational energy to cause this process to take place. The electrons and protons will disappear, and neutrons will appear in their place.

Neutron stars are very complex. Astrophysicists think that they contain neutrons, electrons, atomic nuclei, and other particles. It is even possible that their cores might be made up of tightly packed quarks. However, most of the matter they contain exists in the form of neutrons, and this is what gives neutron stars their character. Since there are not enough electrons left to resist collapse, the matter that makes up a neutron star becomes even more condensed than that of a white dwarf. Where a white dwarf has a diameter about a hundred times smaller than our sun's, a neutron star is about seven hundred times smaller than a white dwarf.

As dense as it is, a neutron star is not the most compressed object possible. If a collapsed star has a mass that is more than about three times that of the sun, gravity becomes so intense that the neutrons cannot resist further contraction. And once this contraction begins, nothing can resist it. If the contraction produces resisting pressures, this will speed up the contraction process, not halt it. According to Einstein's general theory of relativity, the existence of pressure only causes gravitational forces to become greater. All of the matter that makes up the star will be compressed into a volume much smaller than an atomic nucleus, and a black hole will be formed.

A black hole is an object so dense that nothing, not even light, can escape from it. When a great deal of matter is squeezed into a very small volume, gravity becomes so strong that neither matter nor light can escape it once it is caught in the black hole's gravitational clutches. And, according to calculations based on Einstein's general theory of relativity, the force of gravity becomes infinite in the center of a black hole.

SUPERNOVA EXPLOSIONS

Before I discuss the properties of black holes in detail, it might not be a bad idea to say more about how such objects are formed. The stars that become black holes do not die in the relatively sedate manner that is characteristic of average-sized stars like our sun. The death throes of a massive star are much more dramatic. When the star's nuclear fuel becomes exhausted, the star's core collapses to form a rapidly spinning neutron star. The energy generated by this gravitational collapse expels the outer layers of the star into space, in a massive explosion. Supernova explosions are quite dramatic events. Some supernovas become 200 million times brighter than our sun when the light they emit reaches peak intensity. If such an explosion took place in a star system thirty light years away, the resulting supernova would, for a brief period, appear to be sixteen times as bright as a full moon. Although this would be a dramatic sight, it would not be a pleasant one to behold, for the radiation that accompanied the light would cause mass extinctions, and most likely bring human civilization to an end.

Supernova explosions are complicated processes, and all the details are not understood. However, astronomical observations have associated neutron stars with supernovas. For example, a neutron star is in the center of the Crab Nebula, a cloud of luminous material that is a remnant of a supernova explosion recorded by Chinese astronomers in 1054. Nearby stars make the hot gases that were thrown off in the explosion visible, creating a nebulosity. The neutron star is too dark to be seen in optical telescopes, but the pulses of the radio waves it emits can easily be detected. Observations made with radio telescopes have established that it is still rotating rapidly.

SPHERICAL CHICKENS

In his book *The God Particle*, Nobel laureate Leon Lederman mentions a joke common among engineers. If a physicist wanted to study a chicken, he would begin by making the statement "Assume the chicken is spherical."

When physicists study any system, they typically begin by making simplifying assumptions. Then, after the equations that describe the simplified version of the system are solved, more complicated cases can be considered. For example, if an astronomer wanted to calculate the path that a comet would follow when it entered the solar system, he would first consider only the gravitational attraction between the comet and the sun. The perturbations caused by the gravitational attraction of the various planets could be added in later.

Frequently, a great deal can be learned by considering relatively simple problems. For example, the first solutions ever found to the equations of relativity turned out to describe black holes. This discovery was made not by Einstein but by the German astronomer Karl Schwarzschild. Schwarzschild was in the German army on the Russian front when he heard of Einstein's work late in 1915. Without neglecting his military duties, Schwarzschild found mathematical solutions to describe the gravitational fields around a point mass. He then found solutions for the fields around any spherical mass. A few months later, in May 1919, he died at the age of forty-two of a rare skin disease that he contracted while serving in the military.

Astronomical bodies are not exactly spherical. The earth bulges outward a bit at the equator, and is slightly flattened at the Poles. However, even though Schwarzschild's solutions were "spherical chickens," the assumptions they depended on came close enough to reality to render them useful. They made it possible to calculate the gravitational fields surrounding large gravitating bodies such as the sun. Furthermore, they reduced to Newton's inverse square law in cases where gravity was not too intense. This came as no surprise. Einstein did not overthrow Newton's law of gravitation when he discovered general relativity. Newton's law gives the correct results when conditions are not extreme, and astronomers use it for most of their calculations. Even inside the sun, relativistic calculations produce results that differ from those given by Newtonian theory only by a factor of about one part in a hundred thousand. One could say that Schwarzschild proved that the Newtonian law of gravitation was part of general relativity.

Schwarzschild produced yet another result, which certainly must have seemed odd at the time. He showed that if the matter in a gravitating body were compressed into a very small volume, then space would become so strongly curved that this body would be cut off from the rest of the universe. Nothing, not even light, would be able to escape from it.

But, in 1916 no one dreamed that white dwarfs and neutron stars, much less black holes, really existed. In fact, sixteen years were to pass before the neutron was even discovered. Consequently, little attention was paid to Schwarzschild's results. Scientists simply assumed that such compressed states were not possible.

Matters did not change much when, in 1939, Robert Oppenheimer and his student George Volkoff published a paper showing that if a neutron star had enough mass, gravity would become so strong that nothing could prevent further collapse. That same year, Oppenheimer published a second paper with another student, Hartland Snyder, describing black hole collapse in detail. But this paper, too, was soon forgotten. Many of the nation's physicists soon became involved in war work, with other concerns on their minds. In any case, most of the scientists who read Oppenheimer's papers tended to believe that stars would probably eject enough mass, in one way or another, to avoid passing over the Oppenheimer-Volkoff limit. Some of them may also have been influenced by a paper written by Einstein, also published in 1939. In this paper, Einstein attempted to show that black holes could never form. He didn't use the term black hole. Nearly two decades would pass before that name was invented. Of course, this conclusion was wrong. Einstein's mathematical calculations were accurate, as far as they went, but some of his assumptions were incorrect. In particular, he did not properly account for the dynamics of the collapsing material.

AN UNBELIEVER REPENTS

Many physicists can trace their "lineages" quite a distance back. For example, Richard Feynman was a student of the American physicist John Wheeler, who spent two years at Bohr's institute of theoretical physics in Copenhagen. Bohr worked with both J. J. Thomson and Ernest Rutherford. Rutherford worked with Thomson at one time, and Thomson was Lord Rayleigh's successor as director of the Cavendish Laboratory. Rayleigh, in turn, was the successor of the great James Clerk Maxwell.

But it is not my intention to construct lengthy genealogies. I merely want to introduce the reader to John Wheeler, one of the most eminent physicists of our time. In 1957, Wheeler began to investigate the behavior of very condensed stars. Although his work confirmed that stars with more than a certain mass must collapse into black holes, he also assumed that stars must somehow eject enough mass to avoid this fate. In 1958, at a scientific conference in Brussels, he suggested that the extra mass would probably be converted into radiation of one kind or another and that the critical state would never be reached. Oppenheimer, who was in the audience, asked whether it wouldn't be simpler to assume that a black hole was formed. Wheeler wasn't swayed. For a number of years, he continued to maintain that mechanisms probably existed that would prevent black hole formation.

Yet it was Wheeler who coined the term black hole in 1967. Earlier that year, the first neutron star was discovered by a Cambridge Ph.D. student, Jocelyn Bell, and her professor, the English astronomer Anthony Hewish. It was now apparent that highly collapsed stars did exist. And once a neutron star had been observed, the idea that black holes might be real suddenly began to seem more reasonable. Wheeler soon became a believer.

Wheeler's choice of the term black hole was a fortunate one. If he had used different terminology, black holes might never have become a topic of such intense public interest.* For example, the Soviet physicists of the time referred (for reasons that will soon become apparent) to black holes as "frozen stars." It is difficult to imagine the lay public becoming excited about objects with a name that conjures up images of Christmas cookies.

CYGNUS X-1

A hundred million or more black holes may exist in our galaxy. Most of them will never be observed. After all, it is not possible to see something that emits no light or any other kind of radiation. However, some black

---

* The term was resisted at first by French physicists. In French, "black hole" has obscene connotations.

holes are very bright. If a black hole is a component of a binary star system, and if it is close enough to its companion, it will continually draw matter from the surface of the other star. As that matter spirals into the black hole, it will rapidly accelerate and become very hot, reaching temperatures of approximately 100 million degrees. This hot matter will emit copious quantities of radiation. Once the matter enters the black hole, any radiation it emits will no longer be seen, though it will produce quite a large quantity before it disappears.

In 1964, astronomers discovered that an object in the constellation Cygnus was emitting large quantities of X rays. At the time, they were unable to study this X-ray source in great detail because X rays are absorbed by the earth's atmosphere. They had to depend on observations made by equipment sent aloft in high-altitude rockets. Rockets, however, were not exactly ideal observation platforms. They remained aloft for too short a time and they were constantly moving.

The data became much better in 1970, when it became possible to use satellites to observe X-ray sources. These observations made it possible to pinpoint the location of the X-ray source—now known as Cygnus X1—more accurately. It was soon established that they were coming from a system that contained a blue supergiant star, known only as HDE 226868, which had an invisible companion.

Today, the great majority of astronomers believe that unless something is seriously wrong with modern theories of stellar collapse an invisible companion must be a black hole. A collapsed star should contract into a black hole if its mass is greater than three solar masses. Using the most conservative assumptions possible, calculations will indicate that the invisible object must have a mass at least 3.3 times greater than that of the sun. This is greater than the black hole limit. And using the most likely figures for the mass of the blue supergiant and its distance from its companion, one finds that the invisible body probably has a mass nine to fifteen times greater than that of our sun.

Since 1970, a number of X-ray sources that are quite likely to be black holes have been found, and few astronomers continue to entertain doubts that black holes really exist. In fact, evidence shows that huge black holes with masses hundreds of millions of times greater than that of our sun exist in the centers of many galaxies, including our own. At one time it was possible to argue that dying stars might always find some way to avoid black hole collapse. Today that idea is no longer tenable.

BLACK HOLE COLLAPSE

As we saw in chapter 6, Einstein's special theory of relativity tells us there are two things that remain the same for any observer in any state of motion: the velocity of light and the laws of physics. As a consequence, certain other things, including space and time, seem to change. Different observers see these from different perspectives. From our point of view, every object in a spaceship approaching the speed of light would seem to contract in the direction of the ship's motion, and time on the ship would appear to slow down. To an observer on the ship, on the other hand, everything would seem perfectly normal. From that observer's point of view, it would be our time that was slowing down, and we would be the ones who were shrinking.

The effects predicted by the general theory of relativity are similar. The general theory tells us that intense gravity can cause distortions of space and time—or of space-time, if we employ the terminology that physicists commonly use. Since space and time are so bound up with each other in relativity, it is convenient to consider them together. There is nothing really arcane about the concept. Einsteinian relativity has three dimensions of space and one of time. It would be perfectly possible to speak of Newtonian space-time if there were any motivation for doing so.

The slowing down of time in gravitational fields is one of the general relativistic effects that has been observed. If we could place a clock on the surface of the sun, it would seem to run more slowly than a similar clock on the earth. This effect has even been observed in the relatively weak gravity of the earth. Very accurate experiments have been performed to show that physical processes run a little more slowly on the earth's surface than they do at some higher altitude, where gravity is weaker.

Thus if we could actually watch a collapsed star contract into a black hole, we should not expect to see the same things as an imaginary observer positioned on the surface of the star. The intense gravitational forces in the reference frame of the second observer cause matters to seem much different.

Before I explain what a black hole collapse would look like, it is necessary to say a little more about the results that Schwarzschild obtained in 1916. As we saw previously, Schwarzschild showed that if all the mass of a body were compressed within a sufficiently small volume, it would become a black hole. He also showed that this volume could be calculated precisely.

Black holes are the remnants of collapsed stars that have undergone supernova explosions. They form when the infalling matter reaches the Schwarzschild radius. Until that happens, all that exists is a collapsing neutron star. From the standpoint of a distant observer, this never happens. To this observer, as the star contracts and gravitational forces increase, time appears to run ever more slowly. The surface of the star approaches the Schwarzschild radius more and more slowly. It never reaches that radius because that is the point at which time stops altogether; you could say that at the Schwarzschild radius, it appears to run infinitely slowly. It is not surprising that Russian scientists described such an object as a "frozen star."

Nor can a distant observer ever see an object enter a black hole. As that object approaches the black hole's event horizon—an imaginary sphere at the Schwarzschild radius—its time slows down, too. It gradually approaches the event horizon—and appears to remain poised there for all eternity.

GRAVITATIONAL REDSHIFTS

I have been speaking of the things that a distant observer would "see." In reality, he would never be able to observe such events. The reason has to do with something called the gravitational redshift. It is possible to gain some insight into the nature of this phenomenon by looking at the different ways in which particles of matter and light lose energy when they escape from the gravitational fields that surround our sun.

The sun is constantly expelling streams of particles that flow outward through the solar system. This phenomenon is called the solar wind. As the particles that make up the solar wind fly outward from the sun, they lose energy. The gravitational attraction of the sun has a retarding effect, and this causes the particles to slow down.

The sun's gravity will also cause a ray of light to lose energy. But since light always travels at a constant velocity, it must give up energy in a different way. Nature has contrived a simple way to make this happen. The wavelength of a ray of light that is climbing out of the sun's gravitational "well" is slightly lengthened. Light at the blue end of the spectrum has more energy and shorter wavelengths than light at the red end. So if the wavelength of light is lengthened, if the light becomes a little redder, its energy will decrease.

In the case of the sun, the gravitational redshift is not large. But in the enormous gravitational fields that surround a collapsing star, they are enormous indeed. A distant observer watching a collapsing star would at first see its light become redder and redder. Then, as the gravitational collapse continued, the redshift would change the light into infrared radiation, and then into radio waves, which are forms of electromagnetic radiation that have longer wavelengths than those of visible light.

As the collapse continued, the wavelengths would become progressively longer, and their energy less and less. At the event horizon, the gravitational redshift would become infinite. Wavelengths would become infinitely long, and their energy would drop to zero. To the outside observer, it would seem that the collapsing star was emitting less and less light at every moment, until it became completely black.

INFINITY AS AN ILLUSION

As we observed previously, when theories in physics predict the existence of infinite quantities, either this is a sign something is wrong with the theory, or that something is impossible. The ultraviolet catastrophe told physicists something was wrong with their ideas about the manner in which radiation was emitted from blackbodies, and thus led to Planck's quantum theory. The prediction that a body moving at the speed of light would have infinite mass told Einstein that no material object could ever attain that velocity.

In the case of a black hole, matters become quite confusing. From the point of view of a distant observer, not only does time seem to run infinitely slowly at the event horizon, we also find that the surface of the collapsing star emits electromagnetic radiation of infinitely long wavelength and zero energy when it reaches the Schwarzschild radius, and that it will remain "frozen" there for all eternity. Surprisingly, none of these things is a problem from the standpoint of an observer positioned on the surface of the collapsing star. Everything happens in a perfectly normal way. There is no freezing at the event horizon, and the star ever more rapidly continues to collapse.

Such an observer would notice nothing unusual as he fell through the event horizon. In fact, unless he had performed measurements that told him in advance exactly where the horizon was, he wouldn't know that anything significant had happened. Of course, only an imaginary observer could make such a journey. If a real one did it, and somehow managed to avoid being killed by the intense radiation present in that environment, he would be torn apart by the intense gravitational forces of the black hole. If he fell into the hole feet first, the gravitational forces on his feet, which would be closer to the masses that were pulling him downward, would be much stronger than the forces on his head. This would cause him to be drawn out into a long, thin thread, a process that some physicists have called "spaghettification." This thread (the observer) would then continue to be drawn toward the center of the black hole.

From the standpoint of a distant observer, it takes an infinite amount of time for a neutron star to collapse into a black hole. From the standpoint of an observer on the surface of the star, this process would take about a day. The gravitational forces produced when the star collapses are so intense that the time perspectives of the two observers are radically different. Nothing is inherently paradoxical about this. An analogous distortion of perspective can be seen in something as ordinary as a map of the earth.

We have all noticed that in maps that use the common Mercator projection land masses that occupy positions at far northern latitudes appear to be much larger than they really are. The island of Greenland seems to be many times its real size. There is no way to avoid distortions when making a two-dimensional map; the spherical surface of the earth cannot be mapped onto a flat surface with accuracy.

You may also have noticed that maps that make use of the Mercator projection do not include the North and South Poles. They are always cut off at some northern and southern latitudes. The reason is that if they were continued all the way to the Poles, the distortion would become infinite. At the Poles, a single point would be mapped onto a line of finite length. The length, of course, would be equal to the width of the map.

It turns out that something similar happens in the case of a black hole. If it appears to a distant observer that a collapsing star becomes frozen, this is only an illusion. In 1958, the American physicist David Finkelstein discovered it was possible to create a reference frame to view the collapse of a star from a standpoint that included the viewpoints of a distant observer *and* of an observer riding on the surface of a collapsing star. In Finkelstein's reference frame, the star did collapse into a black hole; there was no infinite slowing of time. It was as though someone had replaced a Mercator projection—with its infinite distortion at the North Pole—with a globe. It seems that even distant observers must agree that black hole collapse really does take place. The reason they cannot see it happen is that gravitational redshifts rob light of all its energy when the event horizon is reached. There is no way to receive signals that will tell them what has happened.

INFINITE DENSITY

If we can find a way to cope with the infinite distortion of the time perspective, there are other infinite quantities that are still troublesome. The Schwarzschild solution to the equations of general relativity provides a realistic description of the process of stellar collapse. After the surface of the star passes the event horizon, the collapse must proceed faster and faster until all of the star matter is compressed into a volume of zero dimensions called the singularity. At the singularity, both the density of matter and gravitational forces become infinite. The matter that once constituted a star is compressed into a mathematical point.

The Schwarzschild solution of the equations of general relativity does not take rotation into account, which means it gives a somewhat unrealistic picture of black hole formation. Almost all astronomical objects spin on their axes. The earth rotates on its axis every twenty-four hours. Our moon turns around every time it makes a revolution around the earth, so that it always turns the same side toward us. At one time, the moon spun more rapidly; its rotation has been slowed by tidal forces. The sun rotates on its axis. Even galaxies rotate. And of course, neutron stars rotate, too.

As I pointed out previously, the equations of relativity are quite complicated and difficult to solve. Schwarzschild was able to solve them because he made a number of simplifying assumptions. Nearly half a century would pass before a solution was found for the case in which there was rotation. This feat was accomplished by the New Zealand mathematician Roy Kerr in 1963. It was later proved that the Kerr solution could be applied to black hole collapse, and physicists now make a distinction between the non-rotating "Schwarzschild black hole" and the spinning "Kerr black hole." The former is a mathematical idealization. It is not likely that such objects exist in nature.

The discovery of the Kerr solution raised the question of whether singularities really formed. Could the rotation somehow prevent all the matter in a black hole from being compressed into a point? For that matter, suppose the collapsing star were not exactly spherical? Could this cause the particles collapsing toward the center of a black hole to somehow "miss" one another, so that they were confined within a small but finite volume of space?

These questions were answered by the British mathematician Roger Penrose in 1965. He proved a theorem showing that if the equations of general relativity were correct, a singularity would inevitably be created. Penrose was not able to show that all matter inside a black hole would be compressed into some zero volume, but he was able to prove there would be a singularity at which gravitational forces became infinite.

DO SINGULARITIES REALLY EXIST?

Mathematicians encounter singularities all the time. For example, the simple quantity $1/x$ becomes singular when the number x is equal to 0. The closer x is to 0, the smaller the denominator becomes. When it is exactly equal to 0, the expression becomes infinite. But no one is particularly bothered by such things. Here, one is dealing only with a mathematical abstraction.

When a physical theory predicts infinite quantities, on the other hand, matters are very different. In particular, the prediction that singularities exist within black holes can be interpreted in only two ways. Either a singularity is a place where space and time cease to exist or it is a place where the theory that predicts their existence breaks down. In either case, the conclusion must be that at a singularity, the known law of physics come to an end.

When a particle reaches a singularity, it is subjected to infinite gravitational and tidal forces. It is pulled apart and crushed out of existence at the same time. There is no place it can go "after" it arrives at a singularity, for it has simultaneously reached the edge of space and arrived at the end of time. At one time there was speculation that particles reaching a singularity could pass through a wormhole and emerge in some other universe. However, it was later proved this was impossible. Such wormholes, if they were created, would pinch themselves off immediately, so that nothing could pass through them.

But this is not the only possible interpretation. It is known that a point exists at which the general theory of relativity breaks down. In order to describe the events that take place in regions near a singularity, it would be necessary to incorporate quantum mechanics into the theory. In other words, create a theory of quantum gravity. But, as I pointed out previously, no one knows how to do this. The force of gravity cannot be renormalized.

General relativity ceases to be valid at distances of 10^{-33} centimeters or less. This quantity is known as the Planck length. It is the distance at which (unknown) quantum effects become so large that Einstein's theory could not possibly continue to be valid. Now, 10^{-33} centimeters is a very small quantity, about 10^{21} (1 billion trillion) times smaller than the dimensions of an atomic nucleus. However, a region with dimensions of 10^{-33} centimeters is still infinitely larger than a point. So some physicists feel comfortable with the idea that all the matter in a black hole is compressed into a region of about this size. Of course, whether speaking of a singularity with dimensions of 10^{-33} centimeters, or one of zero volume, the other properties of the black hole remain the same: there is a singularity (or a near-singularity) and an event horizon. There is nothing in between.

Superstring theory represents an attempt to unify all four forces. So, if a successful theory were found, we would automatically have a theory of quantum gravity. Since this is not likely to happen for decades, if ever, we can only speculate as to what conditions in a region with dimensions of the order of 10^{-33} centimeters might be like. It may be that space-time ceases to exist at this level. It has been pointed out that there is no reason to be compelled to believe that space and time are the most fundamental quantities. Even if this is not the case, we are likely to find that events in this "Planck region" bear even less resemblance to those of the everyday world than those in the worlds of quantum mechanics and relativity.

ONE INFINITY REMAINS

If a theory of quantum gravity were developed, it is unlikely it would contradict the conclusion that from the standpoint of a distant observer, time runs more and more slowly as the surface of a collapsing star falls toward the event horizon, or the conclusion that in our frame of reference, there are no true black holes, only "frozen stars" in which massive amounts of stellar material are piled up on the event horizons. In our frame of reference, black holes take an infinite amount of time to form.

This conclusion is not contradicted by observations of systems such as Cygnus X-1. The matter being drawn toward such a black hole would behave in exactly the same manner whether the hole's mass were on the event horizon or inside it. Newton showed centuries ago that any spherical body will attract other bodies as though all its mass were concentrated at its center. It isn't necessary to use general relativity to describe the manner in which this infalling matter will behave. In this case, as in many other circumstances, gravity is still weak enough that Newton's law remains valid. And it is certainly easier to use!

TRAVEL TO OTHER UNIVERSES

I probably should not leave the subject of black holes without first discussing this topic, even though it has little or nothing to do with the idea of infinity as it applies to black holes. Too much has been written about space travel to allow me to ignore it.

When physicists studied the properties of Kerr black holes in detail, they found it appeared possible that particles falling into such a hole might miss the singularity entirely. The mathematical models they used implied that these particles would then travel to another universe. And if this happened, it was possible for the particles to go from this second universe to a third. In fact, this was the kind of journey that could be repeated endlessly.

This gave rise to the idea that it might be possible to undertake such a journey through one or another of the massive black holes that presumably exist at the centers of galaxies. Since these black holes are massive, they have Schwarzschild radii that are correspondingly large. Since the event horizon is a great distance from the central singularity, tidal forces are much less intense than they are around a black hole that has formed from a collapsed star. There is no reason why a spaceship could not enter such an object while leaving its passengers intact.

Of course, it would be a one-way journey. A spaceship could no more leave the black hole than anything else could. But perhaps it could follow a trajectory prescribed by the Kerr solution and wind up in an alternate universe. For that matter, perhaps it could make a series of such journeys and return to its own universe at some other point in space and time.

Although a great deal of speculation has been heard on this subject, it probably adds up to nothing but a theoretical fantasy. It is true that the

Kerr solution to the equations of general relativity takes the rotation of a black hole into account. But, like the Schwarzschild solution, it does make use of certain mathematical idealizations. No mathematician or physicist has yet been able to solve the equations of general relativity in the general case; that is, in such a way that solutions could be found for any possible problem. Some simplifying assumptions are always made. Thus if the Schwarzschild solution can be likened to a spherical chicken, then the Kerr solution might be described as a chicken with an ellipsoidal body and a spherical head.

9

∎●

IS THE UNIVERSE FINITE, INFINITE, OR IMAGINARY?

IS THE UNIVERSE finite or infinite? Opinion on this subject has wavered back and forth for nearly 2,500 years. The idea that the universe is infinite, and that it contained innumerable worlds, was advanced by some of the pre-Socratic Greek philosophers, notably Democritus. This doctrine was "refuted" by Aristotle, who maintained that the world was finite and that, furthermore, no real infinite quantities existed. As we saw earlier, Aristotle taught that there could be only potential infinities. Aristotle may have thought he had settled the question, but the idea of an infinite universe refused to die. It cropped up again in the thought of the Greek philosopher Epicurus, who adopted Democritus's system as the basis for his own philosophy, and in *On the Nature of Things*, a book-length poem by the Roman poet Lucretius in which Epicurus's ideas are expounded.

During the Middle Ages, the growing authority of Aristotle caused the idea of a finite universe to become part of Church doctrine. Thus when Bruno, who had read Lucretius and some of the pre-Socratics, developed his doctrine of the plurality of worlds, the idea was judged heretical. But it was not long before the pendulum swung back the other way. By the seventeenth century, the idea of an infinite had become commonplace.

The same kinds of vacillation can be seen in the scientific thought of the twentieth century. At times scientists have believed that the universe was finite, and at others they have leaned in the opposite direction. In fact, at one point some very compelling empirical evidence favored an infinite universe. But then opinions shifted again, and as I write this, a great many physicists and astronomers—the majority perhaps—believe that our universe is so close to the borderline we will never know whether or not the universe is infinite.

Yes, there is a borderline. According to Einstein's general theory of relativity, the average curvature of space in our universe can be either positive or negative. Positive curvature corresponds to a finite universe, which closes in upon itself. Negative curvature produces a universe that is infinite. There is also a third possibility, a universe in which the average curvature of space is zero. Such a universe would also be infinite, but just barely so. It may sound strange to speak of a universe that is "just barely" infinite. However, any quantity of positive curvature, however small, produces a universe that is closed and finite. A universe with zero average curvature really is on the borderline.

I'll discuss the Einsteinian conception of the universe in more detail shortly. In particular, I'll explain what positive and negative curvature are, and discuss the properties of finite and infinite universes in detail. But it will first be necessary to backtrack a little and look at the situation in which the science of astronomy found itself during the early years of the twentieth century.

MEASURING THE DISTANCES OF STARS

By around 1910, astronomers had observed and cataloged many thousands of stars. By studying the light the stars emitted, they learned a great deal about the chemical composition of stellar atmospheres. Yet in most cases, they had no way of telling how far away a star was from the earth. A great number of stars were concentrated in the Milky Way, a band of light that arches across the sky, but lacking the ability to make distance measurements, astronomers were unable to determine either the exact structure or size of this huge stellar conglomeration. At the time, only one method for measuring the distance of stars existed, the method of stellar parallax. If a star was close enough to the earth, its position would seem to shift slightly as the earth traveled around the sun. The distance of the star could then be found by triangulation.

However, the shifts in star position were small. For example, the Alpha Centauri system is closer to Earth than any other star system. It is only a little more than four light years away. But four light years is a huge distance by terrestrial standards. A light year is approximately 6 trillion miles. The apparent position of Alpha Centauri shifts by only about three-quarters of one second of arc, or about one five-thousandth of a degree, over a period of six months. This shift is approximately equal to the diameter of a dime when viewed at a distance of two miles. An angle of this size can be measured. However, the size of the stellar parallax diminishes rapidly for stars that are progressively farther away. Today, astronomers can use this method to measure the distances of stars that are a few hundred light years away, which is not much by astronomical standards. For example, the earth is about 30,000 light years from the center of our galaxy, which is about 100,000 light years across.

During the early twentieth century, the stellar parallax method was about ten times less accurate than it is today. As a result, astronomers could determine the distances of no more than a few hundred stars. If a star exhibited no parallax, its distance remained unknown. Measuring its apparent brightness didn't help. If a star appeared brighter than others in its vicinity, this could mean either that it really was brighter than they were, or simply that it was closer to Earth. In effect, astronomers could only study the stars as though they were part of a two-dimensional array. Except for a few near stars, they were deprived of a third dimension.

Because astronomers possessed only one distance yardstick of limited usefulness, they were unable to determine the nature of the nebulae, the bright patches of light observed among the stars. Some of these nebulae had an irregular appearance, while others possessed a spiral structure. As long ago as the mid-eighteenth century, Immanuel Kant had suggested that the spiral nebulae were huge systems of stars. However, during the initial decades of the twentieth century, most astronomers believed the nebulae were whirlpools of gas, relatively near Earth, in which new stars were being formed. If astronomers had known how far away the spiral nebulae were, they would have realized they had to be massive collections of stars; otherwise they could not produce enough light to be seen from Earth.

EARLY TWENTIETH-CENTURY "COMPUTERS"

Nowadays, astronomy is quite a sophisticated affair. Professional astronomers rarely make visual observations. Instead, the light that passes through telescopes is collected by electronic detectors that feed the resulting data directly into computers. In some ways, the methods that existed at the beginning of the century were surprisingly similar. In fact, astronomers then had their "computers," too.

Astronomy in the early twentieth century was mostly a matter of collecting data. Many thousands of stars were observed, and data about them was entered in catalogs and star charts. A lot of tedious work needed to be done, and most university astronomy departments had rooms full of employees whose job it was to enter star names and seemingly endless columns of numbers on blank catalog pages. At Harvard College Observatory, these employees were mostly women. Their boss, astronomer Edward Pickering, paid them twenty-five cents an hour, and called them "computers." Pickering was eventually to amass a collection of about a quarter of a million astronomical photographs, and the resulting data would fill seventy volumes of the Harvard Observatory *Annals*.

In 1912, one of Pickering's "computers" made a discovery that was to transform the science of astronomy. Henrietta Leavitt had been assigned the task of finding variable stars; that is, stars whose brightness varied in time. While performing this task, Leavitt noticed a relationship between the intrinsic brightness of a certain type of variable star, called a Cepheid variable, and its period, the time between two points of peak luminosity. The brighter a star was, the longer its period. This implied that if astronomers could measure the distance to a few nearby Cepheids, they would know the distance to them all. After all, if a star's intrinsic brightness is known, its distance can be computed from its apparent brightness. This method is a little like computing the distance of a lightbulb by measuring the light it emits. If one doesn't know the wattage of the bulb, no conclusions can be drawn. But if one knows that the bulb is emitting, say, 75 watts of energy, the calculation is an easy one, provided of course that some apparatus for measuring the light is available.

THE REALM OF THE NEBULAE

Using the new technique, astronomers were suddenly able to measure the distance of almost any star, including stars that were not Cepheids. If one member of a binary star system is a Cepheid, then the measurement of its period tells astronomers how far away its companion is, too. If they know how far away its companion is, they know its intrinsic brightness, and also that of other stars like it. Furthermore, it so happens that our galaxy is surrounded by groups of stars called globular clusters. They are far enough away to assume that all the stars in a cluster lie at about the same distance from Earth. If the distance of one star in the cluster is known, that of all the other stars is known as well.

Stars can be classified according to their color, which is related to their surface temperature. For example, Epsilon Orionis is a blue giant with a temperature of about 25,000° C. It is classified as a B0 star. Our sun is a medium-sized, yellowish star. It has a surface temperature of about 6,000° C, and is classified as a G2 star. All the stars in a given class will have about the same intrinsic luminosity. Hence, once the distance of one star in a class is known, the distances of the others can be computed. It is only necessary to measure their apparent brightness as seen from Earth. And if the distances of individual stars are known, it is possible to determine the size and shape of groups of stars, such as our galaxy.

SHAPELY AND HUBBLE

Using methods based on knowledge of the intrinsic brightness of stars, the American astronomer Harlow Shapely was able to estimate the size of the Milky Way and determine that our sun occupied a position between the Milky Way's center and the galactic rim. Shapely's figure for the diameter of our galaxy, 250,000 light years, was about two and a half times larger than the one that is accepted today. However, even a figure containing this large an error represented an advance. Astronomers had previously had no way to obtain any kind of estimate of the Milky Way's size. For all they knew, it could have had a diameter of no more than a few thousand light years.

Shapely believed that the spiral nebulae lay outside our galaxy. If this was true, then it followed that the spiral nebulae must be bodies comparable in size to the Milky Way. In other words—using a term that became common during the early decades of the century—they were other "island universes." Shapely could not prove his theory, however. At the time, no telescope was powerful enough to distinguish individual stars in distant galaxies. Consequently, the idea that the spiral nebulae were clouds of gas within our own galaxy could not be disproved.

The task was accomplished by the American astronomer Edwin Hubble. Not exactly beloved by his contemporaries, Hubble had an aloof demeanor and was sometimes impolite. Born in Missouri, he spoke with an Oxford accent that seemed to many an affectation, even though he had indeed studied at Oxford, as a Rhodes scholar. Shapely, for one, was convinced that if Hubble were awakened in the middle of the night, he would speak with a Missouri accent.

Hubble was one of the great astronomers of the twentieth century. In 1919, after he returned from military service in World War I, he went to California to work with the recently completed 100-inch telescope at Mount Wilson. He strongly suspected that the spiral nebulae were galaxies, and he intended to prove it.

Hubble established first that the nebulae of irregular shape did indeed lie inside the Milky Way and were what most astronomers had thought them to be: clouds of gas. Hubble then turned to the spirals. He found Cepheid variables in the M33 nebula in 1926 and in the Andromeda nebula in 1928. Not only did this show that they weren't gas clouds, it also became possible to compute the distances of these objects; it immediately became apparent they were farther away than anyone had imagined, which was a dramatic discovery indeed. In 1910—as far as anyone knew—the universe could have been no more than a few thousand, or tens of thousands, light years across. Now it was apparent that it contained galaxies millions of light years away.

THE EXPANSION OF THE UNIVERSE

In the course of studying the galaxies, Hubble discovered that most of them were moving rapidly with respect to Earth and to the Milky Way galaxy. He was able to determine their velocities by means of a phenomenon known as the Doppler effect. In 1842, the Austrian physicist Christian Doppler worked out a relationship between the perceived frequency of a sound and the motion of the object emitting that sound. He had noticed that the apparent pitch of a locomotive whistle was higher when a train was approaching an observer than when it was moving away.

Doppler's explanation was relatively simple. Sound is a wave phenomenon. Like an ocean wave, a sound wave consists of a succession of crests and troughs. The crests correspond to regions in which air molecules are squeezed together, and the troughs to regions in which the air is slightly more rarefied. Sound, in other words, is nothing more than a wave of compression that travels through the air.

When an object is moving toward an observer, it is a little closer every time it produces a wave crest. This causes the crests to be bunched closer together. The time intervals between successive crests are shorter. This corresponds to a higher tone. On the other hand, when the object is moving away, the opposite happens; the distance between successive crests is longer, and the perceived tone is lower.

Light, which is also a wave phenomenon, also exhibits a Doppler effect. When an object that is emitting light moves rapidly away from an observer, wavelengths become longer and are shifted toward the red end of the spectrum. When an object is moving rapidly toward an observer, wavelengths become shorter and the shift is to the blue.

Galaxies that move rapidly away from the earth do not take on a red appearance. In fact, their visual appearance remains pretty much the same. As some of the visible light they emit reddens, invisible ultraviolet light is redshifted into the visible part of the spectrum. Nevertheless, the redshift or blueshift of light from a galaxy is something that can be precisely measured. It is only necessary to examine certain characteristic wavelengths associated with certain elements characteristically found in stellar atmospheres and then determine how much they are shifted. By studying the light emitted from a distance object, astronomers can determine its velocity, and its chemical composition as well.

When Hubble studied the Doppler shifts of the light from the galaxies, he found that while a few, such as the Andromeda galaxy,* were moving toward the earth, the great majority were moving rapidly away. Furthermore, a simple relationship existed between distance and velocity. If galaxy B was three times as distant as galaxy A, it typically receded at a velocity that was three times as great. And if galaxy C was ten times farther away, it moved ten times as fast.

Naturally, this didn't imply that the Milky Way was the center of the universe and that everything was trying to get away from it as quickly as possible. The galaxies seemed to be moving away from Earth because they were all moving away from one another. One can compare the universe to a lump of rising raisin bread dough. As the loaf expands, the distances between all the raisins increase. If you select a raisin at one end of the loaf as a reference point, it is the raisins on the opposite end that seem to move away the fastest. Of course, like all analogies, this one breaks down at a certain point. A loaf of raisin bread has edges, but—as we shall see—the universe has no boundaries.

Hubble spoke of receding galaxies. Today we know that galaxies are generally bound together in clusters, and in clusters of clusters of galaxies. This does not really complicate the situation much. Think of the raisins in the loaf of bread as clusters rather than individual galaxies. All the basic ideas remain the same.

Don't imagine some force exists that causes galaxies and clusters of galaxies to recede from one another. The recession is caused by the fact

---

* Andromeda is a member of the "local group" of galaxies, a small cluster of about two dozen galaxies of varying sizes that are bound together by gravity. These galaxies orbit around one another in complicated ways.

that the volume of space is expanding. At the risk of oversimplifying a bit, as time passes, the objects that make up the universe maintain the same relative positions in a larger space. The reason this is an over simplification is that groups of galaxies have motions unrelated to the expansion of the universe. However, the larger the distance scale considered, the less important local motions become. Our Milky Way galaxy and the other members of the local group, are moving through space at a velocity of 600 kilometers per second. But 600 kilometers per second is not a very large quantity compared with the recession velocity of very distant galaxies, which can be a considerable fraction of the speed of light.

EINSTEIN'S GREATEST BLUNDER

After he published his general theory of relativity in 1916, Einstein set to work to apply his theory to the universe as a whole. He discovered that his equations implied that the universe had to be either expanding or contracting. Quite naturally, he was startled by this result. Since ancient times, it had generally been agreed that the universe was static. Aristotle had taught that the universe was unchanging. Copernicus, Galileo, and Newton had viewed it in the same manner. Early twentieth-century astronomers agreed. Anything but a static universe was out of the question.

Einstein often questioned commonly accepted scientific ideas, but in this case, he did not attempt to overturn prevailing opinion. Instead, he looked for a way to create a theoretical description of a static universe. Noticing that his equations allowed, but did not require, the inclusion of a certain term, Einstein added it, calling it the "cosmological constant." This term, as it turned out, had a simple physical interpretation. The cosmological constant corresponded to a repulsive force that would balance out the force of gravity at large distances. True, it would have to be a very strange force, unlike any other known to physics. Once it was added, however, it gave the result Einstein wanted: a universe that did not change over the course of time.

The only problem was that Einstein's theory really didn't work. In

1922, the Russian mathematician Alexander Friedmann showed that, even with the cosmological constant included, Einstein's equations described a universe that could either be expanding or contracting. When he published his results, he pointed out that Einstein's assumption that the universe was static was only an assumption; it was not supported by observations. Einstein's initial reaction was one of skepticism; Friedmann must have made a mistake, he decided. But after examining the paper more thoroughly, he had to agree the calculations were correct. Nevertheless, he continued to maintain that the universe remained the same over long periods of time.

Then, in 1927, the Belgian mathematician Georges Lemaître discovered that if Einstein's theory was correct, the universe could *not* be static. Einstein's universe was unstable. The slightest disturbance would upset the balance between gravitational attraction and cosmic repulsion. A static universe was like a pencil balanced on its point; it was poised to begin moving one way or the other at any moment. A static state simply could not be maintained.

Finally, in 1929, Hubble announced his discovery of the recession of the galaxies. Empirical evidence indicated that the universe was expanding. Einstein finally gave up on the cosmological constant two years later, calling its introduction the greatest blunder of his career.

OPEN AND CLOSED UNIVERSES

If the universe is expanding, it raises the question of whether it will always continue to expand or whether the gravitational attraction between the matter it contains will eventually bring the expansion to a halt. After all, every particle of matter in the universe attracts every other particle. Admittedly, gravity is a weak force. For example, if I pick up a steel object with a magnet, the magnet exerts a greater force on it than the entire earth. However, the universe contains a great deal of matter and the cumulative effects are quite large. If the expansion of the universe is not coming to a halt, it must at least be slowing down.

If we know how rapidly the universe is expanding, then it is possible to calculate what the mass density of the universe must be if the expansion were to be halted. This mass density is approximately equal to 5 x 10^{-30} grams per cubic centimeter, or about three hydrogen atoms per cubic yard. Unfortunately, the mass density of the universe cannot be measured directly. Scientists are able to obtain good estimates only of the quantity of mass contained in stars, and only a small fraction of the matter in the universe—something between 1 and 10 percent—exists in star form. At present, it is not known how much nonluminous matter there is.

If the mass is greater than the critical density, then the equations of general relativity imply not only that the expansion of the universe must eventually stop, but also that the universe is finite in extent. If there is enough mass to eventually halt the expansion, then there is enough mass

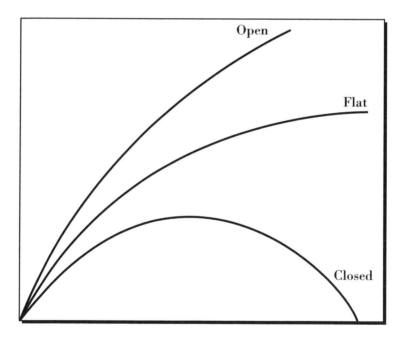

Figure 8. An open universe expands forever. The average distance between galaxies always continues to increase. The expansion of a flat universe also continues forever, but its rate of expansion eventually slows to almost nothing. The expansion of a closed universe eventually comes to a halt. It then enters into a state of contraction. It finally collapses upon itself in a big crunch.

to cause space to curve in upon itself. A universe of this character is said to be closed. The average curvature of space in a closed universe is positive. In a way, space in a closed universe is analogous to the spherical surface of Earth. If such a universe lasted long enough, a spaceship could—at least in theory—start out in one direction and reach its starting point from the other direction after traveling for billions of years. But there is an important difference between such a universe and the earth. The earth's two-dimensional surface is curved in a third dimension, but there is no fourth spatial dimension in which the space of a closed universe can be curved. Do not try to visualize what a closed universe looks like. Even theoretical scientists cannot do that. Instead, they work with Einstein's equations, which give a clear picture of the structure of space-time under such circumstances.

If a closed universe has a "shape" that is somewhat analogous to that of a sphere, the curvature of space in an open universe, one in which the mass density is less than the critical figure, can be compared to that of a saddle. A saddle curves upward in the front-to-back direction and downward from side to side. If the saddle curved downward in one direction at the same rate that it curved upward in the other direction, the ends would never meet. Thus an open universe is infinite. In such a case, space is said to have a negative curvature.

One more possibility is a universe in which the mass density is exactly equal to the critical figure and the average curvature of space is zero. Such a universe can be compared to a flat plane that extends outward in every direction. Like an open universe, a flat universe is infinite. In such a universe, the rate of expansion grows slower and slower, approaching zero, but never quite getting there.

THE BIG BANG

If the universe is expanding, there must have been a time when it existed in an extremely compressed state. In fact, a rough estimate of the time when the universe began can be obtained by measuring the expansion rate today. But the calculation of a date of birth for the universe involves a number of uncertainties. For example, as I write this the rate of expansion of the universe is not known precisely. Different astronomers have come up with different figures and agreement has not yet been reached. Furthermore, the age of the universe also depends upon the amount of matter it contains. A larger mass density means there was more gravitational retardation in the past, and this affects the calculation of the universe's age. It is reasonably certain, however, that the universe began between 10 billion and 20 billion years ago; an age of 15 billion years is often cited as a good compromise figure.

Evidence exists that the universe began in a big bang. About 15 billion years ago, shortly after the big bang, the universe was a hot, glowing fireball filled with rapidly moving subatomic particles. Three important kinds of evidence support this conclusion. One, the present expansion of the universe, has been mentioned. The second is the existence of a cosmic microwave background. In 1964, two Bell Laboratories physicists, Arno Penzias and Robert Wilson, discovered that a strange kind of radiation was falling on the earth from every region of the sky. This radiation consisted of radio waves in the microwave part of the spectrum and was similar to that emitted by a blackbody at a temperature of about three Celsius degrees above absolute zero, or 3 K. Here, K stands for Kelvins, the temperature scale named after the nineteenth-century British physicist Lord Kelvin. The only difference between the Celsius and Kelvin scales is that they have different zero points: 0 K is equal to $-273°$ C.

Since Penzias and Wilson made their discovery, the microwave background has been measured over and over again with increasing precision. It is now known that it is equivalent to that radiated by a blackbody at a temperature of 2.75 K, and it does not vary according to the season or time of day. Blackbodies radiate different amounts of energy at different temperatures, and no differences between the background and the spectrum of a theoretical blackbody have ever been found. The intensity of the microwave background is slightly higher in one direction of the sky than it is in the other. This is attributed to the motion of Earth and the Milky Way through space.

When blackbody radiation is observed, scientists are in effect seeing the big bang fireball. The reason they see radio waves rather than light is that the radiation was emitted 15 billion years ago. Since that time it has undergone an enormous redshift. The wavelengths of what was originally visible light have been lengthened to radio waves. Another way of looking at this phenomenon is to imagine the light waves being "stretched" as the universe expanded. As the universe grew larger, the distance between each successive wave peak became greater, just as the distance between clusters of galaxies became greater.

It is this effect that gives the solution of Olbers' paradox (discussed in chapter 3). When scientists look into the sky in any direction, they do see "light" everywhere. But this light has dimmed in intensity over a period of billions of years so that all that can be observed today is a dying ember.

Finally, the reason the microwave background is seen everywhere is that the big bang was something that happened everywhere. It is wrong to think of the big bang as an explosion in which matter was hurled into a larger preexisting space. There was no space outside the big bang. The universe was once much smaller than it is now, but there was nothing outside it. Then, as now, it had no boundaries. We live inside the space in which the big bang originally took place. Of course, today that space is greatly expanded.

MEASURING HELIUM

The third kind of evidence to indicate there was a big bang about 15 billion years ago is the fact that the universe is about 25 percent helium by weight. The rest is mostly hydrogen. Such elements as carbon, nitrogen, oxygen, silicon, and iron, which are so common on Earth, contribute only a small fraction of the total mass of the universe. One could say that our planet and the organisms that live on it are nothing but cosmic impurities.

The percentages of hydrogen and helium do not vary much from place to place in the universe. Stars are about 25 percent helium. Astronomers find roughly the same quantity in young stars, in old stars, and in those brightly glowing objects known as quasars, which existed many billions of years ago, and whose light can still be seen today. Interstellar gas clouds are about 25 percent helium. Helium is also found in cosmic rays. The majority of the cosmic ray particles are protons. Others consist of nuclei of heavier atoms. When cosmic rays are studied in detail, helium nuclei are present in about the same ratio they are found everywhere else.

It is not hard to explain why there should be so much hydrogen in the universe. A hydrogen atom consists of one protein and one electron, two of the three basic particles of which matter is made. Helium, on the other hand, is more complex. A helium nucleus consists of two neutrons and two protons. To explain a 25 percent helium abundance, it is necessary to see how such a quantity could have been formed.

Hydrogen is converted into helium in stars. The light and heat that stars produce are products of nuclear fusion reactions of various kinds. The most common such reaction is one that converts four protons into a helium nucleus. This is a multi-step reaction, similar to the one that takes place in a hydrogen bomb explosion. The reason that four protons can be converted into a nucleus containing both protons and neutrons is that at one point in the reaction a proton is converted into a positron and a neutron. This is the opposite of the beta decay process in which a neutron decays into a positron and an electron. Yes, a proton can become a neutron just as a neutron can become a proton, providing that all the accounts (for example, energy and electric charge) balance. Since a neutron and a positron have the same +1 electric charge that a proton possesses, everything balances out just fine.

It is possible to calculate approximately how much hydrogen has been converted into helium in stars during the last 15 billion years. When this calculation is performed, a figure of much less than 25 percent is obtained, which forces us to accept the conclusion that most of the helium observed must have been made somewhere else.

The helium could easily have been created in the big bang fireball. Calculations indicate that when the universe was less than about one minute old, the temperatures were too high for helium nuclei to exist. If any had been formed they would have been torn apart by collisions with other particles. But after one minute, temperatures would have decreased to a point where helium nuclei would be stable. Collisions still took place, but they were not so disruptive; the average energy of the sub-atomic particles was smaller. Then, as the universe continued to expand, temperatures dropped further and matter became more dispersed. Helium production slowed down, and the process eventually came to a halt.

Admittedly, this argument has a certain circularity. We have to assume a big bang happened in order to explain the observed quantities of helium in the universe. In turn, this confirms the hypothesis that a big bang happened. But the truth is that a very plausible account of helium production has been created, and there seems to be no other way to explain why so much of this element exists. Furthermore, a body of related evidence strengthens the case. Deuterium, an isotope of hydrogen, is present in our universe in a concentration of about fifteen parts per million by weight. This is important because deuterium cannot be manufactured in stars. It can only be created under very special conditions, such as those that presumably existed in the big bang.

A deuterium nucleus consists of one proton and one neutron. It is said to be an isotope of hydrogen because its chemical behavior is similar. Atoms of both hydrogen and deuterium have nuclei that contain a single positive charge that is circled by a single electron. The proton and the neutron in the deuterium nucleus are very weakly bound. If such a nucleus was created in a stellar interior, it would break up almost as soon as it was formed. The first energetic particle that happened by would knock them apart. On the other hand, a small amount of deuterium could easily have been created in the big bang, at about the same time that helium was formed. Under such conditions, particle collisions would have been frequent enough to create the isotope, but temperatures would have been low enough to allow at least some of the deuterium to survive.

THE FATE OF THE UNIVERSE

All the evidence points to a big bang. But it does not tell us what will happen to the universe in the future. If the universe is open, it will expand forever. After billions of years, all the existing stars will burn out or collapse into black holes. For a while, the universe will contain enough interstellar gas to create new stars, but supplies of this gas will eventually be exhausted, and the universe will become a cold dark space. Even after this happens, the expansion will continue relentlessly so that matter is progressively more dispersed.

On the other hand, if the universe is closed, the expansion will eventually stop, and a phase of contraction will set in. Once this contraction begins, gravitational forces will cause it to proceed ever more rapidly until the universe collapses in a "big crunch" that is the analog of the big bang. A closed universe must eventually come to an end.

For all practical purposes, the third possibility, that of a flat universe, can be neglected. The probability that the universe is in such a state is zero. The mass density of the universe could have an infinite number of different values. But if the universe were flat, this quantity would have to be *exactly* equal to the critical density. In any case, if the universe were flat, its fate would be much like that of an open universe. The only difference would be that the rate of expansion would eventually slow down to almost nothing.

Since it became apparent that the universe can be either open or closed, scientists have tried to find ways to determine what character our universe has. When Einstein published his paper on the static universe in 1917, he assumed that the universe was closed. He seems to have made this choice on aesthetic grounds. In 1917, no evidence would have pointed either to an open or a closed universe. Most likely, Einstein felt that making this assumption would avoid potential problems with infinity.

Hubble was the first person to attempt to settle the question empirically. He reasoned that if the universe were closed, galaxies would seem to be less numerous as one looked out to greater and greater universes. If one looks out to a distance of, say, 100 million light years, the volume of space enclosed within this radius would be less in a closed universe than it would be in one that was open. If there was less space, there would be fewer galaxies. During the early 1930s, Hubble began counting galaxies. Unfortunately, he seems to have reached his conclusion that the universe was closed a bit too hastily, basing it on small discrepancies he had found in his redshift data. Today, it is generally agreed that Hubble proved nothing.

As other astronomers attempted to improve on Hubble's result, it gradually became obvious that it would be difficult, perhaps impossible, to obtain an answer to the question by making astronomical observations. There were just too many uncertainties when galaxies were observed at great distances. And, of course, the mass density of the universe could not be measured directly. As early as 1933 it became obvious the universe contained matter that astronomers could not see. In that year California Institute of Technology astronomer Fritz Zwicky found a large cluster of galaxies in the constellation Coma Berenicies that seemed bound together by the mutual gravitational attraction of the galaxies. But the amount of mass in the galaxies' stars was only a fraction of the amount needed to hold the cluster together. According to Zwicky, the problem was a "missing mass." To use modern terminology, the cluster contained dark matter. Dark matter is just what the words imply—matter that is known or presumed to be present, but which astronomers cannot see. Here, dark really means "invisible." Dark matter is matter that emits no light, or any other kind of observable radiation.

THE CASE FOR AN OPEN UNIVERSE: A BIG HOLE IS FOUND IN THE ARGUMENT

In 1974, four American astrophysicists, J. Richard Gott III, James Gunn, David Schramm, and Beatrice Tinsley presented evidence they claimed indicated that the universe was open. They based their calculations on the quantity of deuterium that was observed. The amount of deuterium that the universe contained depended, they said, on the density of matter at the time it was created. If one knew the density of matter then, it was possible to calculate the density of matter today. The result was a figure considerably less than the critical density.

Since 1974 this calculation has been refined, and similar calculations done for other light isotopes, such as helium 3 (so called because a helium 3 nucleus contains three particles, two protons and a neutron) and lithium 6. The results indicate that the mass density of the universe is no more than 6 percent of the critical value.

Today, scientists don't consider this argument too convincing. Calcu-

lations based on observed concentrations of deuterium and other light elements tell us only how much baryonic matter there is in the universe.
The term baryonic matter means "ordinary matter"; in other words, matter made of neutrons, protons, and electrons. Neutrons and protons be-
long to a class of particles called baryons. Electrons are not baryons, but
they can be left out of the accounting because they are so light. Electrons
make only a small contribution to the universe's mass density.

The argument says nothing about the number of other particles that
could have been produced in the great bang and could still be present in
great numbers today. These particles include those known to exist and
those whose existence has been suggested on theoretical grounds but
have not yet been observed in the laboratory. There is no reason why such
non-baryonic particles could not make a significant contribution to the
mass density of the universe. If they did, the universe could be closed,
even though the contribution of baryonic matter was small.

One fact that has been established is that the universe contains a great
deal of matter that astronomers cannot see. At least 90 percent—possibly 99 percent—of the universe consists of dark matter of one kind or another. It has been established that spiral galaxies, including our own, are
surrounded by dark halos. No one is sure what the halos are made of, but
it is known they are there because their gravitational effects can be observed. It has also been established that additional dark matter is present
in clusters of galaxies. Observations cannot establish whether there is
enough of it to close the universe. But it is certainly a possibility.

NOBODY KNOWS AND YOU CAN'T FIND OUT

If a currently accepted theory of the universe is correct, the universe is
so close to the borderline between open and closed that we will never be
able to determine its true character. In 1980, the American physicist
Alan Guth proposed his inflationary universe theory. According to Guth,
the universe underwent an extremely rapid "inflationary expansion"
when it was a fraction of a second old. In a tiny period of time equal to
about 10^{-32} seconds, its volume may have increased by a factor of 10^{50} or

more. The expansion began when the universe was about 10^{-33} seconds old and ended shortly thereafter. A much slower expansion followed.

It might seem outrageous to speculate about events that took place in 10^{-33} seconds. But if you have confidence that general relativity is a correct theory, then it is indeed possible to consider what might have been happening at such an early time. In fact, it is possible to go back to an even earlier time before the general relativity theory breaks down. It would have to be replaced by a theory of quantum gravity if you wanted to consider what was happening before a time of 10^{-33} seconds.

I won't be discussing the inflationary universe theory in detail here. The interested reader is referred to my book *Cosmic Questions* (which also contains a detailed discussion of dark matter). However, I do want to make two points. First, the inflationary universe theory seems to be the only one capable of explaining why the present-day universe possesses certain observed features. Second, many physicists believe that conditions in the early universe were such that an inflationary expansion of some kind or another was inevitable. The quantum fields that existed in the early universe should have created an enormous outward force. Putting this a little differently, for a brief moment in time there really was a cosmological constant.

I'll give just one example of the inflationary theory's potential for explaining conditions today. Consider the extraordinary evenness of the cosmic microwave background radiation. It is the same in every direction. This radiation was emitted when the universe was about 300,000 years old. Thus it has been traveling through space for 15 billion years. So when we look at opposite sides of the sky, we see regions that are 30 billion light years apart. But what could make them so much alike? With no inflationary expansion, these regions could never have been in contact with one another. After all, the universe is only 15 billion years old, and it would take at least 30 billion years for any causal influence to travel from one region to the other. Nor would there have been time for this to happen in the past. For example, when the universe was one billion years old, these same regions were two billion light years apart.

On the other hand, if inflation did take place, then there was a time when everything that we see was packed together in a relatively small

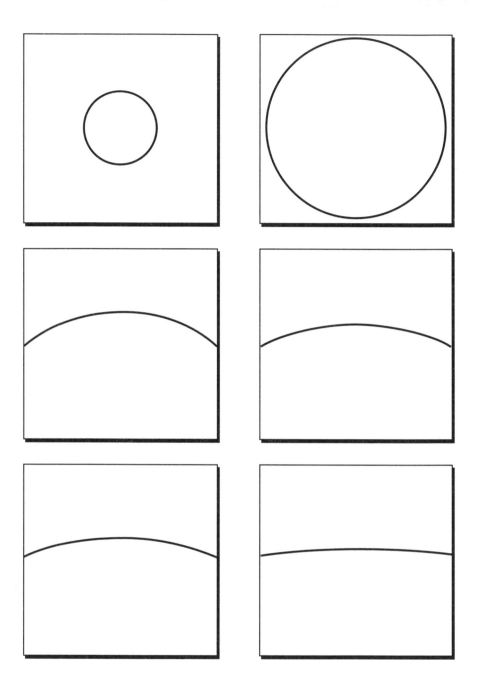

Figure 9. As a curved object expands, its surface gets progressively flatter. The illustrations above represent an expanding sphere. Only part of it is shown in the last four boxes. It is believed that the universe went through an inflationary expansion a tiny fraction of a second after the big bang, and that this expansion "flattened out" the universe to such an extent that the average curvature of space is very close to zero.

space. These widely separated regions could have been in contact. Furthermore, there could have been some "smoothing out" process to ensure that the universe would be uniform now.

The inflationary universe theory is one that cannot really be "proved." In fact, it was originally devised for the specific purpose of explaining why the universe is the way it is; it doesn't yield any testable predictions. However, no other theories can explain what the inflationary universe theory explains. At present, scientists know of no way to explain why the universe has the properties it has without making the assumption that there was a period of inflationary expansion.

If the inflationary universe theory is correct, then the universe must be very nearly flat. The theory predicts that the universe is so close to the borderline that we will never be able to tell whether it is barely open or barely closed. An inflationary expansion has the effect of "flattening out" the curvature of space. In order to see why this should be so, consider the analogy of a balloon. If a balloon contains just a little air, its surface will be highly curved. If it is then inflated, the curvature will become less, and its surface will become flatter and flatter. Of course, when a balloon is blown up, it will increase in size by no more than a factor of ten or twenty, which is almost nothing compared with the factor of about 10^{50}, which is characteristic of the inflationary expansion.

Another example involves comparing a small asteroid to Earth. If one were to stand on a small asteroid that measured one mile across, its surface would appear to be obviously curved. But Earth's surface appears to us to be flat, even though we know that our planet is a roughly spherical body. Again, the difference here is a small one compared with that which would have been produced by an inflationary expansion. The diameter of Earth is only about 8,000 times greater than that of our imaginary asteroid, not 10^{50} times greater.

So perhaps it is not so surprising that no one has ever been able to tell whether our universe is infinite or not.

THE BIG BANG SINGULARITY
AND IMAGINARY TIME

During the 1960s, Stephen Hawking and Roger Penrose proved a series of theorems to show that if the general theory of relativity is correct, then there must have been a big bang singularity. In other words, the theory implies that the universe began in a state of infinite density. If the universe is closed, everything was initially compressed into a single point. Furthermore, a second singularity will be formed when the universe collapses in the big crunch many billions of years from now. On the other hand, if the universe is infinite, it was always infinite. But this does not allow us to avoid the idea of a big bang singularity. In the beginning, the universe and everything in it would have been equally compressed.

Of course, we don't have to take these results literally. Hawking and Penrose began with the assumption that general relativity was always correct. But there is a point at which general relativity breaks down. Thus all that we know is there was initially something resembling the singularity in the center of a black hole. Either it was a true singularity, or it was something very, very small.

Though no one can say whether or not a quantum theory of gravity would imply the existence of a big bang singularity, it is still worthwhile to see whether its prediction can be avoided. All the laws of physics break down at a singularity, after all. If theory says that one existed, then we could never say what happened before the singularity, or describe the manner in which the universe came into existence. For that matter, we would not be able to say whether there was such a thing as a time "before" the big bang, or whether both space and time came into existence at the moment of creation.

Stephen Hawking and the American physicist James Hartle have found a way to avoid the conclusion of a singularity in the big bang. Their hypothesis is not supported by any significant empirical evidence and Hawking has been careful to emphasize that it is only a "proposal." In other words, what Hawking and Hartle have done is to describe one way in which the universe *might* have begun.

The proposal, which is discussed in Hawking's *A Brief History of Time*, makes use of the concept of "imaginary time." It must be emphasized that the word *imaginary* is being used in a technical, mathematical sense, not in the sense in which we ordinarily use it. If there is—or once was—such a thing as imaginary time, it would be just as real as the time we experience in the everyday world.

Imaginary time is related to the concept of an imaginary number. Imaginary numbers were discovered in the sixteenth century when mathematicians began to wonder what the meaning of such quantities as $\sqrt{-4}$ might be. The square root of −4 cannot be a positive number, because any two positive numbers give a positive result when they are multiplied together. Similarly, it cannot be a negative number because a positive number results in this case also. For example, −2 x −2 = 4. So mathematicians invented the imaginary number i, which had the property of i × i = −1. In other words, i was the square root of −1. In this system, the square root of −4 is just 2i. Imaginary numbers can be added, subtracted, multiplied, and divided just like real numbers. They can also be added to real numbers. When this is done, you get what is called a complex number; for example, the quantity $\sqrt{4} + \sqrt{-4}$ is 2 + 2i.

Imaginary numbers are more than a mathematical curiosity. They are used extensively in such fields as physics and engineering. The waves in Schrödinger's versions of quantum mechanics are described by complex numbers. This was why only the square of the wave amplitude (the amplitude of a wave is its height) could be given physical interpretation. The imaginary numbers disappeared when the quantities were squared.

But in the Hawking-Hartle theory, the imaginary numbers do not disappear. Imaginary time is given a real physical interpretation. Here, when time becomes imaginary, it resembles a dimension of space. According to Hawking and Hartle, if you look back far enough in time, then the dimension of time becomes imaginary. In other words, if you look at conditions early in the history of the universe, it is possible that time ceases to be time in the ordinary sense. It becomes something resembling space instead. Today, there are three dimensions of space and one of time. In the Hawking-Hartle theory, there were originally four dimensions of space.

The Hawking-Hartle theory contains no singularity because you cannot look all the way back to time "zero." Time becomes imaginary first. In this theory, the universe has no beginning. A "beginning," after all, is something that happens in time. If their proposal is correct, there originally was no time.

The Hawking-Hartle proposal also implies that if the universe is closed, it will have no end. As the universe collapses in a big crunch, time will again become imaginary, and there will again be only four dimensions of space.

No one knows if the Hawking-Hartle proposal will turn out to be true. The two authors have shown only that their theory is a theoretical possibility. But the theory does allow you to avoid having to deal with a big bang singularity, that infinitely compressed state where time, space, and all the laws of physics come to an end.

The theory of imaginary time also provides a possible explanation of what might happen to the matter that falls into the singularities at the centers of black holes. According to Hawking, these particles may travel through imaginary time to a "baby universe." This odd-sounding term denotes a small, self-contained region of space-time. Since it would contain nothing but the matter that fell into the black hole, it would be quite small compared with our universe. Hawking's baby universes, if they exist, would presumably remain attached to our universe only briefly. According to Hawking, they could reattach themselves at some point. In such cases, they would have the appearance of black holes.

WELL-ESTABLISHED THEORY AND WILD SPECULATION

Books like this one, which discuss the big bang, the inflationary expansion, and such things as Hawking's imaginary time theory together, often give the impression that all these theories are on an equal footing. In truth, some of the ideas discussed in this chapter are considered well established, while others are quite speculative. They most certainly do not have equal probabilities of being true.

For example, the British astrophysicist Michael Rowan-Robinson has estimated the probability as 99 percent that the big bang theory gives a correct description of the universe from a time of one second after the beginning of the big bang. On the other hand, he thinks there is only a 30 percent chance that there was a period of inflationary expansion. Rowan-Robinson goes on to assign a probability of 10 percent to superstring theory. Of the imaginary time theory, he says it has only a 1 percent chance of being "the way forward."

Rowan-Robinson's estimates are on the conservative side. I would guess that most cosmologists would assign more than a 30 percent probability of the inflationary universe theory. Nevertheless, the vast majority of them would agree on some things. No one argues about the fact of the big bang theory; it is considered well established, and most scientists believe it allows us to make deductions about the state of the universe when it was less than one second old. Similarly, most scientists would agree that Hawking's and Hartle's theory is speculative, and that the inflationary universe theory is somewhere in between.

10

INFINITE WORLDS

IN GIORDANO BRUNO'S TIME, to speak of an infinity of worlds was heretical. Today, scientists commonly speak of the possibility that there might be an infinite number of universes. In fact, a number of variations are found on this theme. One is based on the idea: If the big bang can happen once, why can't it happen many, or even an infinite number of times? Another stems from current thinking about quantum mechanics, the basic concept being, if the universe is viewed as a quantum-mechanical system, it is necessary to assume that an infinite number of universes exist to which different probabilities can be assigned. And in quantum mechanics, anything you can assign a probability to "exists" in some sense of the term. Finally, a theory suggests that universes might be able to reproduce themselves through a kind of "budding" process. It is possible that a universe might be able to produce little space-time bubbles, which then break off from the parent universe and undergo inflationary expansions of their own.

To be sure, all these theories are speculative, but speculation is as important a part of science as observation and experiment. Before we can discover what is true, it is necessary to find out what might be true. If it is found that the laws of nature do indeed allow for the existence of other universes, then something has been gained, even if these other universes can never be observed.

If other universes exist, it is not possible to say where they are. The word *where* refers to a location in space, and the space that we observe is something that is contained within our universe. Nevertheless, it is possible that some of these other universes are connected to our universe by microscopic wormholes. If this is the case, these wormholes might affect the character of our universe in certain ways. And if that is the case, then the existence of other universes could be inferred, just as the existence of virtual particles (which cannot be seen, either) is inferred. The possible existence of other universes is a speculative idea. But it is not idle speculation. As we shall see, the creation of theories of "alternate universes" might lead to some real advances in scientific knowledge.

THE UNIVERSE IS A BIG NOTHING

If the inflationary universe theory is correct, then it is likely that our universe originally contained only a very small quantity of matter, and perhaps none at all. Almost all of the universe's mass would have been created during the period of inflationary expansion. As the universe rapidly grew larger during the short inflationary period, matter and energy would have rushed in to fill the rapidly expanding space. The reason this could have happened is that the universe contains a great deal of negative energy that could well balance its positive mass. Since mass and energy can be equated through Einstein's equation $E = mc^2$, it is possible that their sum may be zero.

Many different kinds of energy exist. However, gravitational energy exists in far greater quantities than any other kind. And gravitational energy is always negative. In order to see why this should be so, consider the case of a comet that is so far away from our sun that it experiences very little gravitational attraction. In such a case, the gravitational energy of the system consisting of our sun and the comet would be close to zero. Suppose, next, that some perturbation—perhaps a close encounter with some other astronomical body—causes the comet to begin moving in the direction of the sun. As the comet moves nearer to the sun, it will travel faster and faster. It will gain what physicists call kinetic energy, or energy of motion. But this kinetic energy cannot be created out of nothing; it has to come from somewhere. The only way the comet can gain kinetic energy is by giving up an equal amount of gravitational energy. If this quantity was near zero to begin with, it follows that the comet's gravitational energy will become increasingly negative as it gets closer to the sun.

We can also look at the process in reverse. Suppose we wanted to send a space probe to another star system. It would be necessary to expend energy in order to send the probe out of the solar system. Since the gravitational energy is close to zero when the probe is far from the sun, negative gravitational energy must have been there to begin with.

Any pair of astronomical bodies, whether they are moons, planets, star systems, galaxies, or clusters of galaxies possess negative energy. If they are so far apart that they experience little or no gravitational attraction, then the gravitational energy is essentially zero. But if they are close enough to attract one another, then this energy is negative. The total amount of negative energy must therefore be enormous. We cannot neglect the mutual attraction of galaxies that are millions or billions of light years apart. It is this attraction, after all, that is slowing down the expansion of the universe.

SOMETHING OUT OF NOTHING

The idea that our universe might have come into existence out of nothing was first suggested by the American physicist Edward Tryon in 1969. Noting that the total matter-energy content of the universe could indeed be zero, Tryon suggested that the laws of physics would not be violated if the universe came from nowhere. Since Tryon published his idea, a number of variations on it have appeared. According to one theory, proposed by four Belgian physicists, the universe might have begun with the creation of a single particle-antiparticle pair. According to another, suggested by Rockefeller University physicists Heinz Pagels and David Atkatz, the universe might originally have had a large number of spatial dimensions and no matter. The big bang could have begun when the universe suddenly "crystallized" into its present form. In other words, a spontaneous event that brought about a sudden change in the number of dimensions of space might have created our universe. Although these hypotheses seem different, they are alike in that they begin with the assumption that there was some event that led to the big bang. Such an event could have been a totally random occurrence governed only by the laws of quantum mechanics. Or, as Tryon has put it, our universe may be "simply one of those things which happen from time to time."

The idea of creation out of nothing has been worked out in great detail by Tufts University physicist Alexander Vilenkin. He says that a universe might be created in a manner that resembles the creation of virtual particles. Quantum fluctuations might cause microscopic bubbles of space-time to come into existence. Many of these would disappear almost as soon as they were formed, but some expanding bubbles would be caught up in inflationary expansions. If this happened, a full-fledged universe would emerge.

If Vilenkin's theory is correct, we would have to modify Tryon's remark and say that the creation of new universes is something that happens *all* the time. For that matter, it opens up the possibility that a highly advanced technological civilization might be able to create universes artificially. It isn't obvious why this would be done, since it might be impossible to observe them, but the creation of a universe should certainly be an easier task than the construction of the time machines described by Kip Thorne and his students. The creation of a submicroscopic bubble of space-time also sounds simpler than, say, finding an equally submicroscopic wormhole, enlarging it to macroscopic dimensions, and shoring it up against collapse. After all, finding a wormhole that is too small to be seen does not appear as an easy task. Becoming a little flippant, one could suggest that our universe might even be the result of a botched Physics 201 lab experiment by the otherworldly equivalent of a college sophomore.

If there are many other universes, perhaps an infinity of them, then where are they? Such a question can only be answered by saying, "Everywhere and nowhere." Our universe has no boundaries, and there is nothing "outside" of it. Recall that even a closed, finite universe has no edges. If other universes exist, they do not exist in time and space, at least not in our time and space. They would be "alternate" universes in every sense of the word.

No one really knows whether any of the scenarios for universe creation I have described have any likelihood of being true. A newly created universe would almost certainly have a size of 10^{-33} centimeters or smaller. Present-day physics has no way of describing events that take place on such a level. I must emphasize once again that scientists do not know whether the notions of space and time continue to have meaning at such small dimensions. Until a theory of quantum gravity is found, it will not even be possible to make educated guesses about the processes that take place in the micro-microworld. All that can be said is that the theories that have been developed are not contradicted by anything known to science. But that might just be because in some areas, scientists do not know very much.

ETERNAL INFLATION

The basic idea underlying the inflationary universe theory is a simple one. Current theory in the field of particle physics predicts the existence of numerous different kinds of as-yet undiscovered particles. The reason they have not been seen is that existing particle accelerators are not yet powerful enough to enable scientists to look for them.* If these particles exist, the quantum fields associated with them would have produced huge energy densities, which would have caused inflation in the early universe. Then, as the universe expanded, the fields became less intense, and inflation quickly came to a halt.

Yet, according to the Russian-American physicist Andrei Linde, it is unreasonable to suppose only one inflationary expansion took place. Quantum fluctuations in the early universe would have caused variations in the strength of the quantum fields. In some regions of space, the fields

---

* Among the motivations behind the discontinued Superconducting Supercollider project was the belief that the energies generated in this huge particle accelerator would be sufficient to produce one of these particles, called the Higgs particle.

would have piled up upon one another, producing enough field intensity to spark inflationary expansions. In other regions, the fields would have been relatively weak. Thus in some places inflationary expansion would happen, while other regions of the universe remained very small. If this picture is accurate, our universe could consist of a large number of domains that are very different from one another.

If Linde's theory of "chaotic inflation" is correct, it is unlikely we will ever observe other domains of the universe. The inflationary expansion of our domain would have made it much larger than the observable universe (recall that we cannot see anything more than 15 billion light years away in a 15-billion-year-old universe). And if this is so, neither we nor our descendants will ever be able to observe a domain boundary.

If nothing more than this were in Linde's theory, it would not be too interesting. After all, it tells us little more than that there might be parts of the universe that are unlike our region, but are too distant to be seen. However, Linde's idea of chaotic inflation is only the beginning of his theory. As Linde points out, if this idea is correct, then universes must be able to reproduce themselves. There must be regions within our universe where microscopic fluctuations that cause quantum fields become strong enough to create new inflationary expansions in little bubbles of space-time. These regions would be rare, but they would exist. And if a new inflationary expansion began, then a new universe would be created. We would not be able to see this newly created universe. It would "bud off" from ours, creating a new space-time region of its own. Note that this is not the same process as the one postulated by Vilenkin, in which a universe is created out of nothing. Linde is talking about the creation of new universes within one that already exists.

Linde calls this process "eternal inflation" because it is something that can happen over and over again. If new inflationary bubbles can be created in our universe, then these bubbles will evolve into universes that can have offspring of their own. Nor is there any reason to think that our universe was the first created. According to Linde, it is likely that universes have existed eternally, and that innumerable inflationary expansions have already taken place.

We can't assume that the other universes would resemble ours to any great degree. There is no particular reason to believe that the laws of physics in different universes would be similar. Initially, during the periods in which inflationary expansions were going on, they would always be the same. But as energies fell, physical laws could, according to Linde, take on many different possible forms. For that matter, Linde says, it is possible that quantum fluctuations could sometimes be so strong that the number of dimensions of space and time could change.

WHY IS OUR UNIVERSE
SO USER-FRIENDLY?

At first glance, Linde's theory seems nothing short of bizarre. If it is correct, then an unimaginably large—possibly infinite—number of universes with every imaginable character exist. In most of them, physical laws might be unlike the ones that we observe. Many or most of the other universes might have different dimensionality, and in some of them the dimension of time might not exist at all. Could such a picture possibly be accurate?

Perhaps we will never know the answer to that question. Linde's theory does solve one difficult problem, however, the question of why it is that our universe is so hospitable to life. The fact that life is possible seems to depend upon a whole series of improbable coincidences. If the laws of physics were only slightly different from what they are, life could never have evolved.

For example, in our universe, neutrons are about one-tenth of a percent heavier than protons. As a result, a free neutron will spontaneously decay into a proton and an electron. But if the proton were the heavier particle, no fundamental laws of physics would be violated. A neutron is not a composite particle containing a proton and an electron; the latter are created in the decay process. Neutrons can transform themselves into protons and electrons because extra mass is available.

But if the proton were the heavier particle, then neutrons would be stable, and it would be protons that decayed. Free protons would be transformed into neutrons and positrons. The positrons thus created would eventually be annihilated when they encountered electrons. In this case, the universe would contain little but neutrons and the gamma rays created by the mutual electron-positron annihilations.

If life is to have a chance to evolve, the four forces—the strong, weak, electromagnetic, and gravitational forces—must be precisely balanced. For example, if gravity were weaker it would not have caused the primordial clouds of hydrogen and helium gas that filled the young universe to condense into stars. Furthermore, gravity's braking effect on the expansion of the universe would be less, so the clouds of gas would rapidly disperse. On the other hand, if gravity had been stronger, it might have caused the universe to collapse in a big crunch before life had a chance to develop. Or, if it failed to do that, the increased force of gravity would at least make it easier for stars to collapse into neutron stars and black holes. It is easy to envision a universe in which the stars suffer this fate before life has a chance to develop on any surrounding planets.

If the electromagnetic force that binds atoms and molecules together were too weak, then solids and liquids would not exist. The universe would contain nothing but gas. If it were too strong, the results would be as catastrophic. The strong repulsion between protons would cause the nuclei of many common elements to be less stable. Some of the very common elements would be radioactive. It is not likely that life could exist on earth if oxygen was radioactive and iron was a fissionable material.

It is not just the intrinsic strength that is important in the forces of nature; they must also exist in the right ratios. The example I just gave illustrates this. If the force of electrical repulsion and the strong force that holds atomic nuclei together are not delicately balanced, then either atomic nuclei will break apart too easily or complex nuclei will never form in the first place. For example, if the repulsion between protons were stronger, then nothing but hydrogen would exist. Under such circumstances, the strong force would be able to bind a proton and a neutron together. But as soon as a second proton was added, the mixture would become unstable. Not even helium, the second-lightest of the elements, could be created.

It is easy to conceive of universes in which the various forces are only slightly stronger or weaker than they are in our own. When the properties of these hypothetical universes are examined, we find that, in most cases, they could not possibly contain life. There would be no way for structures of the necessary complexity to evolve. Some of these hypothetical universes turn out to be composed of 100 percent hydrogen or 100 percent helium. In some, stars are blown apart in catastrophic explosions almost as soon as they are created. In others, stars never form in the first place. Some contain nothing but gas; in others, not even hydrogen and helium gas can form. Some are filled with intense radiation. In others, stars collapse into black holes quickly, and everything becomes cold and dark.

A similar observation can be made about the dimensionality of space. It is unlikely that life could ever have evolved if our universe did not possess exactly three spatial dimensions. Stable orbits could not exist if the number were either more or less than this. Orbital stability is necessary if stars are to possess planets. If there were more than three, any planet that happened to form would spiral into the sun. Furthermore, all the stars in a galaxy would spiral into the galactic core, creating an enormous black hole.

HIGHER LEVELS OF COMPLEXITY

If life is to exist, many other seemingly improbable coincidences must exist at higher levels of complexity. For example, the creation of carbon, and of many elements heavier than carbon, seems to depend upon the existence of certain happy accidents. If the nuclei of these atoms didn't "just happen" to have precisely the right properties, these elements would not exist, or would, at best, have been created only in minute quantities.

I must digress a little here on the question of how chemical elements are created. Only hydrogen, helium, and small quantities of such light nuclei as deuterium and lithium were made in the big bang. All of the other elements were created by the nuclear reactions that take place within the interiors of stars. When some of these stars explode as supernovas, such common substances as oxygen, nitrogen, carbon, and iron are spread through space. It would be perfectly accurate to state that

Earth and most of the organisms on it are largely composed of cosmic debris.

The creation of nuclei is a step-by-step process. The lighter elements are made first, and heavier ones are created from them. For example, if a carbon nucleus collides with an alpha particle, then an oxygen nucleus will be formed. A carbon nucleus contains six protons and six neutrons. If the two protons and two neutrons of the alpha particle are added, there are eight protons and eight neutrons in all, and oxygen is the result. Other elements can be formed if carbon or oxygen pick up a stray proton or two.

In general, nuclei made up of alpha particles, or helium nuclei, are stable. Carbon or oxygen are two examples. The carbon nucleus, which contains twelve particles in all, can be thought of as a combination of three alpha particles. Similarly, the oxygen nucleus contains four. Because these nuclei are especially stable, they play an important role in the synthesis of the elements. In particular, if carbon could not be easily made in the interiors of stars, neither carbon nor elements heavier than carbon would exist in significant quantities. Elements upon which life depends, such as nitrogen and oxygen would be absent, and planets such as Earth, which have crusts and cores made of heavy elements, would not exist.

THE BERYLLIUM BOTTLENECK

An exception to the rule that nuclei made of alpha particles are stable is the element beryllium. Like deuterium, beryllium-8, which contains four protons and four neutrons, is easily broken apart. But beryllium is one of the ingredients from which carbon is made. Carbon is created when beryllium-8 collides with an alpha particle. The fact that this happens is something of a miracle. Such a collision should simply break the beryllium nucleus apart, though of course if this did happen, the heavier elements could not be synthesized, and we would not exist to wonder about it.

The reason we are here is that nature has intervened to break open the

beryllium bottleneck. To understand this process you need to know a little nuclear physics. The concepts involved are relatively simple.

In chapter 5, we saw that electrons in atoms had various discrete energy levels, and that a quantum jump between two levels caused the
emission or absorption of a photon of light. As a result, an atom can emit
light only of certain specific wavelengths. Each wavelength corresponds
to a different photon energy, and that energy must be exactly equal to the
difference between two energy states. For example, if an electrical current is passed through a gas, the gas will begin to glow. If the light emitted is passed though a prism and examined in detail, it is found to be
composed of a series of discrete spectral "lines." At most wavelengths,
no light is emitted at all.

Nuclei have energy levels, too. Naturally, these have nothing to do
with the emission of light. But they can determine whether a particular
nuclear reaction will take place. Now it so happens that the carbon nucleus has an energy level of just the right size to absorb the typical energy of collision between an alpha particle and a beryllium nucleus.
Thus, instead of breaking the beryllium nucleus apart, the alpha particle
will be absorbed and a carbon nucleus will be formed.

In 1954, the British astronomer Fred Hoyle pointed out that carbon
had to have a certain energy level if the heavier elements were to be
synthesized in stars. At the time, no one took this idea very seriously, especially nuclear physicists, who were not accustomed to having astronomers intrude upon their field. But when some physicists finally did look
for the level in the hope of getting Hoyle to shut up, they found it at exactly the place he had predicted. Eventually, it dawned on scientists that
the only reason life existed in the universe was that something had
caused one of the energy levels of the carbon nucleus to be very finely
tuned indeed.

THE CURIOUS PROPERTIES OF WATER

Moving up to an even higher level of complexity, we find that the laws of
nature have other special properties, ones that are needed to make life—

at least life as we know it—possible. For example, carbon atoms can bind to one another in chains to form the molecules that are the basis of life. If they could not, amino acids and proteins could not exist. Nor could life exist if those substances did not have certain special chemical properties.

Water is one of the few substances that expands, rather than contracts, when it freezes. If it did not expand, the earth's oceans would have long ago become frozen. Any ice that happened to form would sink to the bottom. Ice at the ocean bottoms could not be melted by the sun, so it would gradually build up. Furthermore, this ice would never melt—much of the sun's energy that is absorbed by oceans and land masses on the earth's surface would be reflected back into space. Eventually, Earth would become a cold, frozen ball.

Water possesses what is known as a high specific heat. The amount of heat required to raise the temperature of water by any given amount is much higher than it is for most substances. This means that as the temperature in an environment goes up or down, the temperature of any body of water will change relatively slowly. Thus water acts as a stabilizing influence. I am very aware of this because I live in San Francisco. The prevailing westerly winds cause the ocean to cool this city in the summer and warm it in the winter. The ocean is why it never snows here. On a larger scale, the earth's oceans help moderate the surface temperatures of our planet. If water did not have a high specific heat, the swings in temperature on Earth would be too great for most, or possibly even all, organisms to endure.

Water has other special properties without which living organisms would find it difficult to survive. One is water's ability to dissolve an unusually large number of different substances, making many of the biochemical reactions upon which life depends possible. Water also has a high surface tension, which is important for the well-being of biological cells. Water has a high heat of vaporization; a lot of energy is required to turn liquid water into vapor. This also has a moderating influence on the environment; for example, lakes do not dry up overnight.

If other universes exist, and the laws of physics and chemistry are different in those universes, life-forms that were not carbon-based, or which did not have metabolisms that required water to function could presumably exist. For that matter, such life-forms might exist somewhere in our own universe. However, it should be clear that whatever life might be like, wherever it might be found, its existence would still depend upon the presence of very special conditions. If life were based on something other than water, it would still most likely need to make use of some substance (which might not even exist in our universe) with properties that were just as unusual. As the example of water indicates, the evolution and continuation of life require environments that have unique properties.

DID GOD STRUCTURE
PHYSICAL LAWS OF OUR UNIVERSE
IN A SPECIAL WAY?

During the nineteenth century, the so-called argument from design was quite popular. In those days, it was fashionable to see God's hand in the wonders of nature and in the diversity of life-forms on our planet. According to the argument from design, God existed because everything was so intricately designed. This argument is unfashionable today. We know, in broad outline, how the earth and living creatures evolved, and feel no need to conclude that divine intervention was involved. However, we cannot help but be reminded of the argument from design when we consider all the improbable coincidences upon which life depends. Indeed, if a person believes in the existence of God, it is tempting to cite these coincidences as evidence.

I believe it is best to keep God out of scientific discussions. However, when we speak of the conditions necessary for life and the possible existence of an infinite number of universes, we have ventured so far into the realm of metaphysics that religious questions pop up quite naturally. I don't intend to take sides on such questions, but I see no reason not to discuss what the implications of such ideas might be for both the believer and the unbeliever. We will find, somewhat surprisingly, that they are really quite similar.

If you are an atheist, then the special character of our universe seems to lead naturally to the conclusion that many, or perhaps an infinite number of universes with different physical laws do indeed exist. If they do, the character of our universe ceases to be surprising. Life would evolve in only a few, very special universes. The vast majority would presumably be lifeless. In other words, our universe has a special character because otherwise we would not be here to wonder about it.

If you believe in God, the possibilities increase. Our universe may indeed have been created to be hospitable to life, but it does not follow that other universes do not exist. In fact, you could resurrect Bruno's argument about the necessity of an infinite number of worlds and say that an infinite God would naturally create an infinite number of universes. I wouldn't venture to comment on the validity of this argument. But at least these days you can suggest it without risking being burned at the stake.

So perhaps the questions about God disappear in the end. Whether God exists or doesn't exist, we still have good reasons for believing that universes might be infinite in number.

QUANTUM COSMOLOGY

I think most scientists will agree with me when I say that quantum mechanics is the most successful theory ever produced in the field of physics. As I pointed out previously, virtually all of modern physics is based upon it, and the predictions of quantum mechanics are confirmed in experiments performed in laboratories around the world every day.

Given the fact that quantum mechanics is *the* basic theory of modern physics, we shouldn't hesitate to apply it to the universe itself. In fact, this is the premise upon which the relatively new field of quantum cosmology is based. Quantum cosmologists attempt to understand what the initial conditions of the universe might have been like, and use these initial conditions to deduce the properties of the universe today. If a theoretical model does indeed produce a universe that looks like ours, this is interpreted to mean that the initial assumptions have a reasonable chance of being correct.

We have already encountered one example of quantum cosmology, the Hawking-Hartle theory of imaginary time. According to that theory, the fact that a universe that had a beginning in imaginary time would evolve into one like the universe we experience makes the imaginary time proposal plausible, even though it doesn't actually prove that the idea is correct.

Although quantum mechanics has been an enormously successful theory, it is also a strange one. Niels Bohr once remarked that anyone who is not shocked by quantum mechanics has not really understood it. Richard Feynman denied even the latter possibility. He characterized quantum mechanics as the theory that "nobody understands." So expect that quantum cosmology will produce ideas that seem strange at first. Again, the concept of imaginary time is a good example.

Quantum mechanics is a theory of probabilities. I gave an example of this in chapter 5 when I discussed the energy levels in a hydrogen atom. With infinite number of such levels, we can speak only of the probability that an electron occupies any given state. In a sense, it is in all energy levels at the same time. Or at least this is the case until an experimental apparatus interacts with the atom, making it do something.

When quantum mechanics is applied to the universe, it is necessary to consider an infinite ensemble of possible universes, to which different probabilities can be assigned. Like the energy levels in an atom, in a sense all of these universes exist simultaneously. Unlike the universes of Linde's theory of chaotic inflation, most of these universes would be quite similar to one another. A given set of initial conditions will produce universes that differ only in small details, such as the distribution of stars or, as Stephen Hawking pointed out, whether or not Madonna's picture appeared on the cover of the last issue of *Cosmopolitan*.

Hawking speaks of different "histories" of the universe. Like the term imaginary time, he uses the word *history* in a technical sense here. History is a reference to an interpretation of quantum mechanics that was developed by Richard Feynman. But we need not concern ourselves with details. Whether one speaks of universes or histories, the main point is that if the fundamental ideas of quantum cosmology are correct, alternate realities coexist with ours. Or, as Stephen Hawking says in his *Black*

Holes and Baby Universes, "We happen to live on one particular history that has certain properties and details. But there are very similar intelligent beings who live on histories that differ in who won the war [He is referring to World War II] and who is Top of the Pops."

Do other quantum universes exist? If they do, could the inhabitants of one communicate with those in another? No one knows. According to physicist Murray Gell-Mann, "This fascinating issue . . . is only now receiving enough attention from theorists of quantum mechanics to be properly studied." It appears that not just an infinity of universes, but an infinity of an infinity of them might exist. If there is any validity to Linde's chaotic inflation and in the quantum cosmologists' concept of different histories of the universe, then the number of universes could be infinite in several different ways.

THE COSMOLOGICAL CONSTANT

As we have seen, if other universes exist, it solves the problem of why our own universe is so hospitable to life. The concept of other universes may answer some other perplexing questions as well; for example, why Einstein's cosmological constant should be so close to zero.

As discussed in chapter 7, Einstein came to regard the introduction of a cosmological constant into his equations as a blunder. Modern scientists are not so sure about this. The most general solutions to the equations of general relativity include the constant. We know that there was in fact a large cosmological constant during the period of inflationary expansion, when there were strong repulsive forces.

Today, the cosmological constant is either zero or so close to zero that it cannot be measured. Otherwise, the forces associated with the constant would affect the motions of distant galaxies, an effect that has not been observed. It is not obvious why this should be the case, as theoretical calculations indicate that the constant should be quite large. There is a huge discrepancy between theory and observation. Calculations indicate that the constant should be 10^{120} times larger than the maximum consistent with astronomical data.

Don't forget, there is no such thing as "empty" space. Virtual particles are constantly being created and destroyed everywhere. The vacuum is filled with the quantum fields associated with these particles. Furthermore, energy, which can be calculated, is associated with all this activity. When this calculation is performed, the self-energy of the vacuum turns out to be enormous. Since mass and energy are equivalent, this energy should give rise to huge gravitational forces. These forces would not vary with distance; on the contrary, they would be the same everywhere. In other words, there would be a cosmological constant.

When Einstein conceived of the cosmological constant, he associated it with repulsive forces. The constant can be either positive or negative, however, and the forces associated with it can be either attractive or repulsive. The forces associated with quantum fluctuations in the vacuum should be so great that the universe should never have been able to expand beyond microscopic dimensions. One could say that accepted scientific theory predicts that the dimensions of the universe should be much smaller than those of an atomic nucleus.

In 1988, the Harvard physicist Sidney Coleman published a paper entitled "Why There Is Nothing Rather than Something" in which he pointed out that if microscopic wormholes connected our universe with an infinite number of other universes, then particles could presumably pass through the wormholes during their brief existence. Coleman calculated that this would have the effect of exactly canceling out the cosmological constant in our universe.

Coleman's results created a great deal of excitement within the theoretical physics community. He had found the first real evidence that other universes might exist. He had certainly not established their reality beyond any doubt. After all, his evidence was highly theoretical in nature. But he had shown that the assumption that other universes were real could lead to a solution to one of the most baffling problems of theoretical physics.

WHY IS A PROTON SO HEAVY
AND AN ELECTRON SO LIGHT?

A proton weighs 1,836 times as much as an electron. Why? Why do protons and electrons possess electrical charges of a particular size and not some other? In general, why do the various subatomic particles have the properties they do?

If you'd asked a physicist such questions in, say, 1970, he would most likely have said something to the effect, "That's just the way things are." The physicists of the 1990s view matters differently. They feel that a complete explanation of the subatomic world will not have been attained until it is known *why* particles have the charges, masses, and other particular properties they are observed to possess.

One of the attractions of superstring theory is the belief that if a successful version of the theory is found, it might very well provide explanations for these things. But it may be that the explanation of the size of charges, masses and so on will come from something other than superstring theory. Stephen Hawking says that particles may acquire certain properties because they are constantly traveling to other universes through wormholes. Hawking points out that if particles are able to disappear into and emerge from wormholes, their masses will be greater than if the particles always remained within the same universe. Furthermore, there would be similar effects on the particles' charge.

If microscopic wormholes exist, they cannot be seen. Thus if an electron traveled through a wormhole to another universe, it would seem to suddenly disappear. Again, this is something that is not observed. Electrons just don't suddenly vanish. But this does not contradict Hawking's theory, which is a theory of particle exchange. According to Hawking, when an electron leaves our universe, a second electron emerges from the wormhole. Neither universe gains or loses an electron; they simply exchange particles with one another.

Every electron is identical to every other electron. They all have the same charge and the same mass. Thus wormhole exchange is a process that cannot be directly observed. As far as we are concerned, an electron was there a moment ago, and it is still there now. There is no way of knowing whether it is the same electron or a different one. However, if calculations were performed that gave the correct values for the electron's properties, such as mass and charge, we would have real evidence that wormholes and other universes existed.

As I write this, such calculations have not been successfully carried out. All I can say is that it is possible to conjecture that particles might acquire charges and masses in this way. Naturally, there are other possibilities. For example, if the laws of physics vary from universe to universe, particle charge and mass might be matters of chance. It could be that in some universes the proton weighs 1,836 times as much as an electron, and in others it weighs five times as much. On the other hand, it could turn out that quantum mechanics and yet-to-be-developed particle theories will tell us that some values of charge and mass are more probable than others. As I write this, speculation about such matters has hardly begun. As a result, it is possible to do little more than cite the different possibilities.

AN INFINITY OF UNIVERSES

Not so very long ago, the concept of alternate realities was encountered only in science fiction. Today, the idea of the existence of an infinity of other universes, some of which may be very different from our own, is an ingredient of respectable scientific thought. As we have seen, no one knows whether such universes really exist, but the assumption they do allows us to deal with problems that would otherwise seem intractable. If there are other universes, the fact that the existence of life seems to depend on so many improbable coincidences seems less puzzling. If there are other universes, it may explain why the cosmological constant is so small. And if there are other universes, we may eventually discover why subatomic particles have certain properties. Scientists may even find out why the physical laws they have been studying for many centuries have the particular character they do.

11

∞

THE MATHEMATICAL SYMBOL ∞ IS, OF COURSE, for infinity. It is a picture of a mathematical curve called the lemniscate, and it was first used to symbolize infinity in John Wallis's *Arithmetic Infinitorum* in 1656. The usage quickly became popular, and the symbol is universally used by scientists and mathematicians today.

The Greek word for infinity was *apeiron*, which means "unbounded." In the ancient Greek world, *apeiron* often took on negative connotations. It could mean "totally disordered," for example; *apeiron* could be used to describe the chaos out of which the world was formed, or a crooked line. According to Aristotle, the quality of being infinite was "a privation, not a perfection." In the world of Aristotle, as in the worlds of his predecessors Pythagoras and Plato, there was no place for infinity.

Aristotle realized that many things in the world seemed to be infinite. Space and time might go on forever, and a line might be composed of an infinite number of points. In order to avoid the disorder he associated with the concept, Aristotle denied that the truly infinite existed. He developed a theory of the potentially infinite, and denied that an infinite void existed outside the celestial spheres.

Aristotle and his predecessor Zeno agreed that infinity could be a problematical matter. It was Zeno who first laid bare the problematical nature of the infinite. By doing this, he believed he could show that the

everyday world of common sense was not the true reality. Few people today would accept the doctrine of Zeno's teacher Parmenides that the
world is an unchanging unity, but the paradoxes Zeno created still fasci-
nate us. Modern philosophers have elaborated upon Zeno's ideas to create paradoxes of their own. One of the simpler ones concerns a lightbulb
that is turned on and off an infinite number of times. As in Zeno's paradoxes, this involves performing an infinite number of acts. Logic tells us
that it will be both on and off in its final state. If the lightbulb is turned
on, then off an infinite number of times, it will be off at the end. But if it
is turned on, and then is alternately turned off and on an infinite number
of times, the final state will be on. This can be represented by the mathematical series:

$$1 - 1 + 1 - 1 + 1 - 1 + \ldots$$

This series seems to have two different sums according to how we group
the numbers. It can be either

$$(1 - 1) + (1 - 1) + (1 - 1) + \ldots$$

which is equal to zero, or

$$1 + (-1 + 1) + (-1 + 1) + (-1 + 1) + \ldots$$

which is equal to 1. Series like these present no problem in mathematics.
Mathematicians call them divergent series and deny that they have any
real sums.* But in the real world a paradox does seem to result if the act
of turning the lightbulb on is represented by the number "1" and the act
of turning it off by the number "−1." It can then be shown that, in the end,
the lightbulb is both in the state 1 (on) and in the state 0 (off).

To the ancient Stoic philosophers, infinity presented no problems when

---

* Series do not have to be ambiguous sums in order to be classified as divergent. Two other
examples of divergent series are $1 + 2 + 3 + 4 + \ldots$ and $1 + \frac{1}{2} + \frac{1}{3} + \frac{1}{4} + \ldots$. Both of these
series increase without limit; in other words, they become infinite. Of course, the first series
becomes large much more quickly than the second.

it was viewed uncritically. The Stoics believed that the world was surrounded by an infinite void, and that time repeated itself endlessly in an infinite series of cycles. They did not think to ask why the world occupied one particular position in the void, and not some other. If they had, they might have realized that to speak of "position" in such a context is not meaningful. After all, if there were a void, if the world were displaced some distance from its present position, nothing would have changed. An infinity of empty space would still be around it. They did not ask how cyclical time began, or how it was possible that there had already been an infinite number of cycles. If they had, they might have realized that infinity was not as simple an idea as they supposed it to be.

Galileo realized there was something baffling about the idea of an infinite universe. His successors found it easy to contemplate this idea, because they didn't engage in reflection about the implications. Because Newton had apparently never thought very much about the infinitely large, he fell into error when he tried to add up all the gravitational forces—an infinite number of them—that would act on an individual star. He mistakenly thought he could show they would balance one another out, and that an infinite universe would be stable. Newton did ponder the infinitely small, but found himself baffled.

Newton's infinitesimals presented mathematical problems, and these problems were eventually solved. In mathematics, problems associated with infinity are usually tractable, however difficult they may initially seem. In physics, on the other hand, the appearance of infinite quantities in a theory is usually a sign that something is terribly wrong. Since modern physics attempts to probe so deeply into the nature of reality, scientists have had to come to terms with the infinite on a number of different occasions. In some cases, they have had only partial success. For example, the theory known as quantum electrodynamics is an extraordinarily successful theory, but no one knows if QED is mathematically consistent. QED views the electron as a point particle. Because the electron's charge is seen as being concentrated in a mathematical point, the "bare" electron turns out to have an infinite charge and an infinite mass that are shielded from our view by infinite numbers of virtual particles. This bizarre picture of the electron is accepted because the consequences of ascribing a finite size to the particle are even worse. When calculations were first made using QED, the quantities the physicists were trying to find turned out to be infinite. So physicists "subtract out" the infinities by making use of the procedure called renormalization. However, renormalization is a mathematically questionable procedure. Thus we have a situation where science uses dubious mathematical techniques in QED and in theories such as quantum chromodynamics (QCD) that are modeled on it, which does raise questions as to how well the fundamental components of nature are understood. Many theoretical physicists hope that superstring theory will eventually resolve the problems, although superstring theory too has been a topic of intense debate. Some scientists think it will eventually lead physics to a kind of theoretical nirvana, while others believe it will eventually prove to be a dead end.

The difficulties encountered by modern physicists show us that the infinite is still as much a mystery as it was in the time of Zeno. To my mind, the infinities encountered in the fields of astrophysics and cosmology are the most fascinating. If we are to believe what Einstein's general theory of relativity tells us, then matter is compressed to infinite density in the interior of black holes. It is agreed that Einstein's theory will break down before this point is reached. But this means only that the true nature of a black hole singularity is unknown. No one is sure whether a singularity is a place where all the known laws of physics break down and where space and time come to an end, or whether a theory of quantum gravity would reveal something totally unexpected.

At present, there is no theory of quantum gravity. Gravity is a more complex force than the other forces of nature. Mass, energy, pressure, and the gravitational fields themselves all give rise to gravitational forces. It has been shown that attempts to renormalize a quantum version of general relativity must fail. The infinities encountered are much worse than the infinities in the theories used to describe the behavior of electrons, quarks, and other particles. Because such a quantum theory of gravity does not exist, scientists are not sure what character space and time might have on a submicroscopic level. Physicist John Wheeler has suggested that, in the region of the very small, space and time might cease to be continuous, and that a kind of quantum "foam" might exist, that space-time might be full of a lot of submicroscopic bridges and holes. Other scientists have suggested that the very concepts of "space" and "time" might cease to have meaning in the region of the very small. Thus, if the density of matter and gravitational forces do not become infinite in the interiors of black holes, one has every right to expect that something extremely bizarre must be happening.

Most of the time, physicists see infinity as a problem that must somehow be eliminated before further progress can be made. But, just as Bruno did in his time, scientists working in the field of quantum cosmology have embraced the infinite, suggesting there really might be an infinite number of quantum universes. In quantum mechanics, we often speak only of probabilities. As soon as we begin to apply the theory to the universe as a whole, we are confronted with an infinite ensemble of universes with different probabilities of existence.

When we play a game such as roulette, we can assign a probability to the likelihood that any given number will appear. This probability is 2.63 percent in American roulette, and 2.70 percent in casinos like Monte Carlo's, which use wheels containing only a single zero (as compared with American wheels, which have both zero and double zero). When the wheel is spun, we know that only one of these probabilities will become real. Two different numbers cannot be the result of a single spin.

But quantum mechanics views the world in a different way. Here, probability is the fundamental concept, and probabilities have a kind of reality they don't possess in the everyday world. When probabilities are assigned to the different positions an electron might have, then we must picture the electron as occupying *all* of those positions simultaneously. When physicists consider the behavior of electrons in atoms, they picture an electron cloud that surrounds the nucleus. Similarly, if an electron can be in some arbitrary number of different energy states, we must say that it occupies all of them. These are not just theoretical models. The existence of such probabilities has been confirmed by experiment, such as when a single neutron has been made to simultaneously follow two different paths, after which it was observed to interact with itself.

Thus if quantum cosmology envisions an infinite ensemble of universes, it is necessary to consider the possibility that there really are other quantum universes, and that many of them might be peopled by beings very much like us. Some scientists, such as Murray Gell-Mann, have even begun to speculate as to whether it might become possible to communicate with these other universes some day.

We also have reason to believe that the number of universes may be infinite in yet another sense. The big bang could be something that has happened not just once, but an infinite number of times. Thus, where some of the pre-Socratic Greek philosophers, and later Bruno, spoke of an infinite number of worlds, we now find respected scientists considering the possibility that there may be an infinity—or perhaps even an infinity of infinities—of universes.

I must emphasize that there is no empirical evidence to indicate that these universes really exist. It is hard to say whether such ideas should be considered as "science" or as a kind of a metaphysical speculation framed in the language of mathematical physics. Nevertheless, the possible existence of innumerable "alternate universes" gives rise to some sobering thoughts. Could it be that there are an infinite number of copies of you and me in other universes? Could our duplicates be living lives that differ from ours in an infinite number of sometimes important, sometimes trivial ways? If universes are created endlessly, does this mean that we—or individuals indistinguishable from us—will live again an infinite number of times?

When I ask such things, I am not engaging in speculation. On the contrary, I am simply relating some of the thoughts that have arisen in my mind, a mind that found itself confronted with the infinite, a mind that feels itself very much in sympathy with Pascal's famous confession: "The eternal silence of these infinite spaces frightens me."

INDEX

Page numbers in italics indicate figures.

Acceleration, 64
Achilles and the Tortoise (paradox), 7–8, 9, 10–11
Adam and Eve, 40, 41
Alaric I, King, 27
Alchemy, 60–61
Alembert, Jean Le Rond d', 66
Alpha Centauri, 160
Alpha particles, 77–78, *78, 79,* 86, 117, *193, 194*
Alternate realities, 198–99, 202
Alternate universes, ix, 157, 188
　possibility of, 208
　theories of, 185
Amino acids, 195
Analyst, The, Or a Discourse addressed to an Infidel Mathematician (Berkeley), 65
Anatomy of Melancholy (Burton), 41
"Anatomy of the World, An" (Donne), 46
Anderson, Carl, 101
Andromeda galaxy, 165
Andromeda nebula, 163

Annalen der Physik (journal), 90
Antineutrons, 103
Antiparticles, 103–04
Antiprotons, 103, 117
Antiquarks, 103
Antoninus, Marcus Aurelius, 24, 25
　Meditations, 25
Aquinas, St. Thomas
　Summa Theologica, 18–19
Archytas of Tarentum, 26
Argument from design, 196
Aristotelian cosmology, 47
Aristotelian thought, 39, 56–57
Aristotle, 19, 23, 26, 32, 33, 57, 158, 166, 203
　time in, 20–21, 22, 25
　proof of existence of God, 18
　refutation of Zeno, 9
Arithmetic Infinitorum (Wallis), 203
"Arrow, The" (Zeno), 59
Assumptions
　in blackbody radiation problem, 84
　erroneous/invalid, 68, 71, 145
　simplifying, 143–44, 153, 157

Astrological signs, 34
Astrology, 27–28, 44
Astronomy, 42–48, 52, 159–64, 175
    Copernican Revolution, 35–38
    development of, 33–35
Astrophysics, 206
Atkatz, David, 186
Atomic catastrophe, 79–82, 89
Atomic collapse, problem of, 85, 87–88
Atomic nucleus(i), 79, 112, 141, 142,
    173, 191, 192
    discovery of, 76–78, 79
Atomic structure, 75–76, 78, 95
Atomic theory, 87–88, 92
    negative reaction to, 88–90
Atoms, 72–73, 74, 75, 80, 81, 118
    behavior of, 91, 98
    light emission, 194
    nuclear model of, 86
    photons' interaction with, 96
Augustine, St., 27–29
    City of God, 27
    Confessions, 28
Augustus, Emperor, 24
"Avatars of the Tortoise" (Borges), 13
Axiom of Archimedes, 69
Axioms, 69
Aztecs, 21

"Baby universes," 182
Baryonic matter, 176
Baryons, 176
Baseball analogy, 1–3, 15–16, 126–27
Becquerel, Antoine-Henri, 74
Bell, Jocelyn, 146
Bentley, Richard, 51
Berkeley, George
    Analyst, The, 65
Berlin Academy of Sciences, 66
Bernoulli, Jakob, 63–64
Bernoulli, Johann, 63–64
Beryllium bottleneck, 193–94

Beta decay, 112, 172
Bible, 33
Big bang, xi, 29, 53, 134, 169–71, 176,
    184, 186, 187, 208
    elements created in, 192
    helium created in, 172–73
    time began in, 20
Big bang singularity
    and imaginary time, 180–82
Big bang theory
    probability of being true, 182–83
"Big crunch," 174, 180, 182, 191
Binary star systems, 147, 162
Biochemical reactions, 195
Black hole collapse, 148–49, 152, 153
Black hole singularity, 180, 182, 206
Black holes, ix, 142, 144, 145, 146–
    48, 151–53, 155–56, 173, 191,
    192, 206, 207
    compression of matter in, 155
    distortion of perspective regarding,
    152
    possibility of travel through, 157
Black Holes and Baby Universes
    (Hawking), 198–99
Blackbody(ies), 81
Blackbody radiation, 82–83, 151,
    170–71
Blackbody radiation problem
    solution to, 84–85
Bohr, Niels, 85–86, 90, 92, 109, 129,
    145, 198
    negative reaction to theory of atom
    of, 88–90
Book of Genesis, 27
Book of the XXIV Philosophers, 33
Borges, Jorge Luis, xi, 13
    "Avatars of the Tortoise," 13
    "Library of Babel, The," 54–55
Born, Max, 92
Brahe, Tycho, 42–43, 45
Brief History of Time, A (Hawking), 181

Broglie, Louis de, 91*n*
Brothers Karamazov, The (Dostoevsky), 29–30
Bruno, Giordano, 38–42, 49, 51, 158, 184, 197, 207, 208
 De l'infinito universo, 40
"Budding," 184, 189
Burton, Robert
 Anatomy of Melancholy, 41

Caesar, Julius, 24
Calculus (the), 51
 foundation for, 67–69
Calculus Made Easy (Thompson), 68–69
Cantor, George, 5–7, 14, 70
Carbon, 171, 192
 energy level, 194
Carbon atoms, 195
Carbon nucleus, 193, 194
Cauchy, Augustin-Louis, 70
 Cours d'analyse, 67–68
Causality, 138
Cavalieri, Bonaventura, 55, 59–60
Cavendish Laboratory, 86, 145
Celestial spheres, 32, 33, 35
Cepheid variable, 161, 162, 163–64
Change, 201, 202
Chaotic inflation, 189, 198, 199
Cheseaux, Jean-Philippe de, 52
Chinese, ancient, 21
Christians/Christianity, 25, 27–28, 30
"Chronology Protection Agency," 139
Cicero, 24–25
 Nature of the Gods, The, 24
Circular time, 20, 21, 27
City of God (Augustine), 27
Coleman, Sidney
 "Why There Is Nothing Rather than Something," 200
Coma Berenicies, 175
Complex numbers, 181

Complexity, higher levels of, 192–93, 194–95
Conduction of Electricity through Gases (Thomson), 86
Confessions (Augustine), 28
Contact (Sagan), 133
Copernican Revolution, 35–38
Copernican system, 44, 45, 46, 47
Copernicus, 35–38, 42, 44, 166
 De Revolutionibus Orbium Coelestium, 35–36
Cosmic cycles, 21–22, 23–24, 26–27, 28, 30, 55
Cosmic microwave background, 170, 171, 177
Cosmic Questions (Morris), 177
Cosmic rays, 101, 172
Cosmic strings, 136
Cosmological constant, 166, 167, 177, 199–200, 202
Cosmology, ix, 33, 46, 206
 Aristotelian, 47
 medieval, 33
 modern, 20
 quantum, ix, 197–99, 207–08
 Stoic, 25–26
Cours d'analyse (Cauchy), 67–68
Crab Nebula, 143
Creation, 26, 28, 29, 30, 40, 180
 out of nothing, 187
Critique of Pure Reason (Kant), 19–20
Curved space, 108, 131–32, 144–45
Cutler, Curt, 136
Cyclical time, 21, 22, 27–28, 29, 205
Cycles, infinite number of, xi
Cygnus X-1, 146–48, 155
Cyrano de Bergerac, Savinien, 41

Dark matter, 175, 176, 177
Darwin, Charles Galton, 86
Davies, Paul, 80
De l'infinito universo (Bruno), 40

De motu corporum in gyrum (Newton), 50

De Revolutionibus Orbium Coelestium (Copernicus), 35–36

Dedekind, Richard, 6

Democratic Rule of Physics, 139

Democritus, 158

Density, 175–76
  infinite, 152–53, 180, 206

Dent, Miss, 97

Derivatives, 67, 68–69

Deser, Stanley, 136

Determinism, 28, 65

Deuterium, 173, 175, 176, 192, 193

Dialectic, 9–10

Dialogue Concerning the Two Chief World Systems (Galileo), 47–48, 49

"Dichotomy, The" (Zeno), 9, 11–12, 14, 67

Differential calculus, 61–64, 65–67

Digges, Thomas, 38, 52

Diogenes the Cynic, 12

Dirac, P.A.M., 96–97, 98, 99–101, 102, 103, 110, 137

Divergent series, 204

Dodgson, Charles Lutwidge (Lewis Carroll)
  "What the Tortoise Said to Achilles," 17–18

Donne, John, 41, 46
  "Anatomy of the World, An," 46

Doppler, Christian, 164

Doppler effect, 164–65

Dostoevsky, Fedor
  Brothers Karamazov, The, 29–30

Dyson, Freeman, 110

$E = mc^2$, 98, 103, 122, 129, 185

Earth, 23, 171, 193, 195
  celestial spheres surrounding, 32, 33
  center of cosmos, 33

center of solar system, 43, 47
galaxies moving away from, 164, 165
geometry of surface of, *132*
motion of, 33–34, 37, 47–48, 153, 170
orbit, 37, 43
surface of, 131, 169, 179

Eastern philosophies, 9

Ecclesiastes, 27

Ehrenfest, Paul, 83

Einstein, Albert, 84–85, 90, 106, 115, 121, 144, 151
  on black holes, 145
  cosmological constant, 199–200
  $E = mc^2$, 98, 103, 129, 185
  general theory of relativity, 31, 113, 114, 131, 137, 142, 154, 159, 166, 169, 206
  greatest blunder of, 166–67
  "On the Electrodynamics of Moving Bodies," 122
  papers on relativity, 125, 132
  and quantum theory, 89
  special theory of relativity, 95–96, 105, 108, 120–21, 122, 148
  and static universe, 174
  theories of relativity, ix, 123, 124, 125, 129

Elasticity, 121

Electric fields, 72, 74, 75

Electric forces, nature of, 111

Electricity, 73, 74

Electrodynamics, 121–23

Electromagnetic force, 112, 191
  theory of, 113, 114–15

Electromagnetic radiation, 73, 81, 83, 150, 151
  interaction with matter, 95–96

Electromagnetic theory/electromagnetism, 72, 75, 81, 87, 122–23

Electromagnetic waves, 85, 124

Electron cloud, 93
Electron orbits, 87–88, 90, 93, 129
Electrons, 77, 86, 101, 102, 137, 190, 191
  behavior of, 98, 206, 207
  as bundles of probabilities, 93–94
  constituent of baryonic matter, 176
  constituent particles of, 118–19
  discovery of, 74–75, 76
  energy levels of, *88*, 194
  energy loss, 79, 80
  have infinite mass, 95–119
  infinite sea of, 99–101
  interaction with protons, 111
  interactions of, 89
  mass of, 75
  number of, in atom, 75, 76
  as point particle, 205–06
  and positrons, 103
  properties of, 201–02
  role in structure of matter, 141–42
  vibrating, 81, 82
Electroweak force, 112
Electroweak theory, 113
"Elementary Exposition of the Higher Calculus" (L'Huiller), 66
Elements, 24, 191
  creation of, 192–93
  properties of, 73
  spectrum of, 89*n*
Ellipses, 37, 47, 50
Elliptical orbits, 43, 45, 62
Energy, 94, 200, 206
  equivalent to mass, 106, 122, 200
  infinite, 127–29
  kinds of, 185–86
  and time, 103–04
Energy loss, problem of, 75, 76, 79, 80, 81
Energy states, 93, 95–96, 99, 104, 194, 207

Epictetus, 24
Epicurus, 158
Epicycles, 35, *36*, 37, 44, 47
Epsilon Orionis, 162
Eternal inflation, 189–90
Eternity, 20
  time and, 28–29
Ether, 124
Euclid
  First Proposition, 17–18
Euclidian geometry, 131
Euler, Leonhard, 65–66
Event horizon, 149, 150, 151, 152, 155, 157
Experimentation, 56–57, 90

Fall (the), 40
Falling bodies, 56–57
al-Fargani, 38
Faster-than-light travel, 120, 126–27, 130, 131
  effects of, 127–29
Feinberg, Gerald, 137, 138
Fermat, Pierre de, 60
Fermi National Accelerator Laboratory (Fermilab), 117
Feynman, Richard, 97, 102, 109–10, 113, 116, 145, 198
Feynman diagrams, *109*, 110
Finkelstein, David, 152
First Cause, 19
First Mover, 18
Flat universe, 169, 174, 179
Fluxions, 61–63, 67
Force of gravity
  see Gravitational force
Forces, 191
  unified, 155
Forces of nature, 112
  theory explaining, 113–15, 116–17
Fractions, 5, 12

Frederick II, king of Denmark, 42
Friedmann, Alexander, 167
"Frozen stars," 146, 149, 151, 152, 155
Furry, Robert, 102
Future, travel to, 127–28

Galaxies, 51–52, 153, 164, 192, 199
  in closed universe, 174, 175
  clusters of, 165–66, 176
  "local group" of, 165n, 166
  mutual attraction of, 186
  recession of, 164–66, 167
  spiral, 176
Galileo, x, xi, 4, 7, 36, 41–42, 46, 61,
    67, 121, 166, 205
  behavior of falling bodies, 56–57, 58
  *Dialogue Concerning the Two Chief
    World Systems*, 47–48, 49
  telescope, 46–48
Gamma rays, 83, 103, 191
Gell-Mann, Murray, 138, 199, 208
General theory of relativity, ix, 31, 113,
    114, 130, 131, 132–33, 136, 142,
    148, 159, 166, 177, 206
  and big bang singularity, 180
  and expansion of universe, 168–69
  law of gravitation part of, 144
  point at which breaks down, 154,
    155, 180
  solutions to equations of, 157, 199
  and time travel, 136–37
Geometry, 60, 131–32
Georgi, Howard, 111
Globular clusters, 162
Gluons, 112
God, 26–27, 28, 29, 60, 61, 196–97
  existence of, 18
  infinite, 32–33, 40
  in infinite space, 51
God Particle, The (Lederman), 143
Gott, J. Richard, 135–37
Gott, J. Richard, III, 175

Grand unified theories (GUTs), 113
Gravitation, law of, 43, 108, 144
Gravitation, theory of, xi, 48, 130
Gravitational energy, 185, 186
Gravitational fields, 144, 150, 206
Gravitational force(s), 62, 112–13, 142,
    154, 191, 200, 205, 206, 207
  infinite, 153, 154
  theory of, 113, 114–15
  in white dwarf, 141, 142
Gravitational redshifts, 149–50, 152
Gravity, 48, 56, 71, 95, 206
  in distortions of spacetime, 148–49,
    151
  and expanding universe, 167
  in infinite universe, 51
  inverse square law, 50, 62, 144
  law of, 49–50, 62
Great year (the), 22–23, 27–28
Greece, ancient, 22
Greek philosophers/philosophy, x, 26,
    28–29, 158, 208
Gregorian calendar, 43
Gunn, James, 175
Guth, Alan, 176–77

Halley, Edmond, 50, 52, 65n
Hartle, James, 180–81, 183
Harvard College Observatory, 161
  Annals, 161
Hawking, Stephen, xi, 97, 116, 136–37,
    139, 180–81, 183, 198–99, 201
  Black Holes and Baby Universes,
    198–99
  Brief History of Time, A, 181
Hawking-Hartle theory, 181–82, 183,
    198
HDE 226868, 147
Heisenberg, Werner, 90–92, 95, 97,
    98, 102, 108
  uncertainty principle, 103–04, 129
Heliocentric theory, 44, 46, 47, 48

Helium, 89, 191, 192
  measuring, 171–73
Helium nuclei, 193
Helium 3, 175
Heresy, 28, 36, 61, 158
  of Bruno, 39–40, 41
Hertz, Heinrich, 72
Hewish, Anthony, 146
Higgs particle, 188*n*
High-energy particle physics, 138
High specific heat, 195
Hilbert, David, 7
Hindus, 9, 21
Hook, Robert, 50
Horoscopes, 33, 44, 45
Hot bodies
  cooling rate of, 63
  emission of light from, 73, 83, 84–85
  energy emitted from, 81
Hoyle, Fred, 194
Hubble, Edwin, 163–64, 165, 167,
  174–75
Hydrogen, 191, 192
  deuterium isotope of, 173
  percentage of, in universe, 171
Hydrogen atom, 75, 76, 80, 93, 172
  behavior of, 92
  energy levels, 198
  quantum theory of, 85–86, 87–90
Hyperreal numbers, 69–70

Imaginary numbers, 181
Imaginary time, xi, 198
  big bang singularity and, 180–82
  probability of, 182–83
Incomprehensible (the), x, 4, 7
Indivisibles, 59, 60, 63
Infinite (the), ix, 13
  in/and modern physics, 72
  prize offered for solution to problem
    of, 66
  problematical nature of, 203–04

Infinite density, 152–53, 180, 206
Infinite energy, 127–29
Infinite mass, 151
  electrons have, 104–06, 107
Infinite number of acts, 11, 12
Infinite numbers, 1–3, 4, 5–6, 7, 14,
  15–16, 20, 70
  defined, 5
  paradoxical property of, x, 3–4
Infinite quantities, 7, 8, 129, 150, 152,
  205–06
  existence of, 158
  removing from QED, 107, 108, 109–
    10, 113, 116
  theories predicting, 71, 150–51,
    154
Infinite regress, 18, 19–20, 127
Infinite time, 17–31
Infinite worlds, 32–53, 184–202
Infinitely great (the), 66
Infinitely large (the), 71, 205
Infinitely small (the), xi, 54–70, 205
Infinitesimals, 59–60, 63–66, 67, 68,
  205
  new mathematical theory of, 69–70
  ratios of, 67–68
  second-order, 64–65
  use of, to perform calculations,
    62
Infinities
  in electron orbits, 87–88
Infinity, 52, 66, 203–08
  as illusion, 150–52
  idea of, x–xii
  paradoxical nature of, 1–16
  in Stoicism, 25–26
  time and, 20
Infinity barrier, 129, 137, 139
Infinity of universes, 202
Inflationary expansion(s), 184, 187,
  188–89, 190, 199
  probability of being true, 182–83

Inflationary universe theory, 176–79, *178*, 185, 188–90
Inquisition, 36, 39–40, 44
Instantaneous velocity, 58–59, 62, 67–68
    defined, 63, 64
Inverse square law, 50, 62, 144
Iron, 171, 191, 192
Iron atom, 75
Island universes, 163

Jackiw, Roman, 136
James, William, 14
Jeans, James, 82
Jordan, Pascual, 97
Judaism, 26–27
Judeo-Christian heritage, 26–27, 28
Jupiter, 32, 46

Kaluza, Theodor, 114
Kaluza-Klein theory, 114–15
Kant, Immanuel, 28, 160
    Critique of Pure Reason, 19–20
Kelvin, Lord, 73, 75–76, 170
Kelvin-Thomson theory, 76
Kepler, Johannes, 42, 43–46, 47, 48, 50
    Somnium, 45
Kepler, Katherine, 44–45
Kerr, Roy, 153
Kerr black holes, 153, 156
Keynes, John Maynard, 60
Kinetic energy, 185–86
Klein, Felix, 69
Klein, Oscar, 114–15
Kronecker, Leopold, 6, 7
Kundera, Milan
    Unbearable Lightness of Being, The, 30

Law of gravitation, 49–50, 62
Laws of nature, 73, 139, 194–95

Laws of physics, 87, 117, 120, 121, 135, 154
    break down at a singularity, 180, 206
    in different universes, 190, 195
    do not change with motion, 148
    end of, 182
    and origin of universe, 186
    and time travel, 139
Leavitt, Henrietta, 161
Lederman, Leon
    God Particle, The, 143
Leibniz, Gottfried Wilhelm von, 51, 63–64, 66
    and differential calculus, 61–62
Lemaître, Georges, 167
Lemniscate, 203
Leptons, 118, 119
L'Huiller, Simon
    "Elementary Exposition of the Higher Calculus," 66
"Library of Babel, The" (Borges), 54–55
Liceti, Fortunio, 49
Life
    conditions necessary for, 196–97
    existence of, 194–95, 202
    and higher levels of complexity, 192–93
    in other universes, 197
    universe hospitable to, 190–92, 199
Life-forms, 195–96
Light
    behavior of, 72
    Doppler effect, 164
    emission of, 84–85
    interaction with matter, 96, 98, 104, 111
    redshift/blueshift, 164–65
    varying with wavelength, 81
    wave and particle character of, 85
    see also Speed of light
Light waves, 72, 122, 171
Light year(s), 53, 71, 72, 160

Limits, theory of, 67–69
Linde, Andrei, 188, 189, 190, 198, 199
Linear time, 26–28, 29, 30
Lippershey, Hans, 46n
Liquids, 141, 191
Literature
  the infinite in, xi, xii, 8
Lithium, 192
Lorentz, Hendrik, 130
Lucretius
  On the Nature of Things, 158

M33 nebula, 163
Magnetic fields, 63, 72, 74, 75, 101
Magnetism, 73, 111
  laws of, 121–22
Marcus Antonius (Marc Antony), 24
Mars, 32, 34
Mass, 201, 202, 206
  equivalent to energy, 106, 122, 200
  infinite, 95–119
  missing, 175
  relativistic increase in, 128, 129
  of white dwarf, 141
Mass density of universe, 168–69, 170, 174, 175, 176
Mathematical formulas, 57, 68
Mathematical idealizations, 157
Mathematical methods/techniques, 59, 61–62
Mathematicians/mathematics, xii, 4, 5–7, 67, 68, 205
  infinitesimals in, 68, 69–70
Mathematics of infinity, x, 5–7
Matrices, 91n
Matrix mechanics, 92
Matter, 73, 85, 94, 186
  creation of, 29
  interaction of electromagnetic radiation with, 95–96
  interaction of light with, 96, 98, 104, 111

nature of, 115, 116, 117
  role of electrons in structure of, 141–42
  wave and particle character of, 91
Mayas, 21
Maxwell, James Clerk, 72, 73, 74, 75, 79, 87, 145
Mechanics, 72
Mechanism, 65
Meditations (Marcus Aurelius), 25
Mental pictures, 98, 102
Mercator projection, 151–52
Mercury, 32
Metaphysics, 196
Michelson, Albert, 123–24
Michelson-Morley experiment, 123–24
Middle Ages, 33
Milky Way galaxy, 52, 121, 159, 164, 165, 166, 170
  size of, 162–63
Milton, John
  Paradise Lost, 41
Mocenigo, Giovanni, 38
Molecules, 95, 195
Montaigne, Michel de, 41
Moon, 32, 33, 34, 43
  rotation of, 153
Morley, Edward, 123–24
Morris, Michael, 133–34
Motion, 18
  accelerated, 57
  impossibility of, x, 9, 59
  and measurement of speed of light, 123–24
  in Newton, 19, 50
  and perspective, 148
  is relative, 121, 124
  in sky, 33–34
  sound and, 164
  uniform/nonuniform, 120–21
Moving bodies, behavior of, 72, 73

Nagaoka, Hantaro, 75
Natural Philosophy of Time, The (Whitrow), 13
Nature, components of, 206
    see also Laws of nature
Nature of the Gods, The (Cicero), 24
Nebulae, 160–61, 162, 163–64
Nebulosity, 143
Negative curvature, 159
Negative energy, 98, 99, 185, 186
Negative-energy electrons, 98, 99–101, 102
Negative mass, 98
Negative numbers, 181
Negative quantities, 98–99
Neutron(s), 77, 101, 118, 141, 142, 145, 172, 173, 190–91
    constituent of baryonic matter, 176
    structure of, 112
Neutron star(s), 141, 142, 145, 191
    collapsing, 149, 151
    discovery of first, 146
    rotation of, 153
    and supernovas, 143
New Testament, 27
Newton, Isaac, xi, 42, 43, 60–61, 67, 69, 97, 121, 156, 166, 205
    argument for infinite universe, 51–52
    De motu corporum in gyrum, 50
    invention of differential calculus, 61–63, 66
    inverse square law, 144
    knighted, 73
    law of gravity, 48, 49–50, 108
    laws of motion, 19
    Opticks, 51
    Philosophiae naturalis principia mathematica, 49, 50, 51, 63
Nietzsche, Friedrich, 30–31
Nitrogen, 171, 192, 193

Nobel Prize, 86, 89, 97, 101, 130
Non-Euclidian geometry, 131–32, *132*
Non-standard analysis, 69–70
Norse mythology, 21
North/South Poles, 152
Nuclear physics, 194
Nuclear reactions, 112, 140, 141–42, 143, 172, 192, 194
Nuclei
    creation of, 193
    energy levels, 194
Number system, 69

Olbers, Heinrich, 52
Olbers' paradox, 52–53, 171
Old Testament, 27
"On the Electrodynamics of Moving Bodies" (Einstein), 122
On the Nature of Things (Lucretius), 158
One (the), 8–11
Oppenheimer, Robert, 102, 145, 146
Oppenheimer-Volkoff limit, 145
Opticks (Newton), 51
Orbital stability, 192
Original sin, 40, 41
Oxford University, 39
Oxygen, 171, 191, 192, 193
Oxygen atom, 75, 76
Oxygen nucleus, 193

Pagels, Heinz, 186
Paradise Lost (Milton), 41
Paradoxes, xii, 28
    Olbers' paradox, 52–53, 171
    with tachyons, 137
    in time travel, 135
    of Zeno, x, 7–8, 9, 10–12, 13–14, 17, 59, 67, 204
Paradoxical nature of infinity, 1–16
Parmenides, 9, 204

Parmenides (Plato), 9, 22

Particle accelerators, 115, 116, 117, 128, 188

Particle exchange, theory of, 201–02

Particle physics, 138, 188

Particle theories
yet-to-be-developed, 202

Particles, 100–01, 102, 103–04, 172–73
as-yet undiscovered, 188
behavior of, 206
charges and masses of, 113
elementary, 118
faster-than-light, 137–38
new, 112
non-baryonic, 176
in superstring theory, 115–16
see also Subatomic particles

Pascal, Blaise, 60, 208

Past (the)
sending messages to, 137, 138
travel to, 127, 128, 135–37

Paul, St., 27

Pauli, Wolfgang, 91, 97, 115

Penrose, Roger, 153, 180

Penzias, Arno, 170

Perspectives, 124–26, 133, 135, 148, 149
distortion of, 151–52

Philosophiae naturalis principia mathematica (Newton), 49, 50, 51, 63

Philosophy, xi, xii, 8, 23, 35

Photon(s), 85, 95, 96, 101
interaction with electrons, 104, 111, 112

Photon energy, 194

Physical laws, 202
structured by God, 196–97
see also Laws of physics

Physics/physicists, ix, 7, 68, 69, 145
birth of modern, 72–75

classical, 90, 129–30
classical vs. quantum, 92–94
infinities predicted by theories in, 71, 205–06
modern, 72
quantum mechanics in, 95
revolution in, 84
see also Theoretical physics/physicists

Pickering, Edward, 161

Planck, Max, 92, 130
quantum theory, 84–85, 87, 151

Planck length, 154

"Planck region," 155

Planck's law, *82*

Planetary motion, 35, 43, 44
book of tables of, 45–46
laws of, 43

Planetary orbits, 37, 43, 47, 50
elliptical, 43, 45, 62

Planets, 33, 34, 191, 192, 193

Plato, 23, 29, 203
dialogues: *Parmenides*, 9, 22
Timaeus, 29

Plum pudding model, 75–76, 77n, 79

Plurality of worlds, 41–42

Poe, Edgar Allan, 53

Poincaré, Henri, 6, 130

Point mass
gravitational field around, 144

Point particle
electron as, 80, 104–06

Points
mathematical, ix
number of, 5–6

Positive charge
in atoms, 76, 77, 78, 79

Positive curvature, 159, 169

Positive-energy electrons, 99, 100, 101

Positive integers, 3, 4, 5, 6, 14, 20–21, 181

Positron(s), 104, 106, 172, 190, 191
  discovery of, 101–03
Possibility of existence
  and assumption of existence, 138
Potential infinities, 20–21, 26, 158,
  203
Probabilities, 184, 198
  in quantum mechanics, 93, 94, 207
Problems (attrib. to Aristotle), 22
Proteins, 195
Proton(s), 77, 80, 100–01, 117, 118,
  141, 142, 172, 173, 190
  constituent of baryonic matter, 176
  properties of, 201–02
  repulsing behavior of, 191
  structure of, 112
Prussian Academy of Science, 85
Pythagoras, 22, 203
Pythagoreans, 22

Quantum chromodynamics (QCD),
  111–12, 206
Quantum cosmology, ix, 197–99,
  207–08
Quantum electrodynamics (QED), 104,
  106–07, 108–10, 119, 205–06
Quantum field theory, 97–100, 104,
  111–12, 116
Quantum fields, 188–89, 200
Quantum "foam," 206–07
Quantum gravity, theory of, 155, 177,
  180, 188, 206–07
Quantum mechanics, ix, 19, 89, 90–
  94, 98, 99, 102, 108, 116, 123,
  129, 154, 202
  applied to the universe, 198
  basic theory of modern physics, 197
  complex numbers in, 181
  and infinite number of universes, 184
  laws of, 186
  probabilities in, 93, 94, 207

and relativity, 97–99
  theory of almost everything, 95–96
  uncertainty principle in, 103–04
Quantum states, 93
Quantum theory, 81, 84, 87, 89, 151
  of hydrogen atom, 85–86, 87–90
Quantum universes, 199, 208
  infinite number of, 207
Quarks, 80, 112, 142, 206
  constituent particles of (proposed),
  117–19
Quasars, 171–72

Radiation, 72
  emitted from collapsed stars, 146,
  147, 151
  theoretical formulas for emission of,
  from blackbodies, 82–83
Radioactivity, 74
Radio waves, 72
Rationalism, 9
Rayleigh, Lord, 81–82, 145
Rayleigh-Jeans law, 82, 83
Reality, 9, 205
Redemption, 27, 40
Redshifts, 149–50, 152, 164–65, 171
Reincarnation, 41
Relativistic effects, 133, 148
Relativistic objects, 136
Relativity, 121
  dimensions of space and time in,
  148
  Michelson-Morley experiment in,
  123–24
  quantum mechanics and, 96, 97–99
  solutions to equations of, 144, 153
Religion, 23, 28
Renaissance, 28, 33, 35
Renormalization, 108–11, 113, 116,
  154, 206
Repulsive forces, 191, 199, 200

Robinson, Abraham, 69

Roentgen, Wilhelm, 73

Roman Catholic Church, 18–19, 33, 39, 40–41, 42, 158
  and concept of time, 28
  and heliocentric theory, 44

Romans, 23, 24–25

Rome, 25, 27

Rotation, 153, 157

Rowan-Robinson, Michael, 183

Royal Society, 50, 60

Rudolph II, Holy Roman Emperor, 44

Russell, Bertrand, 12

Rutherford, Ernest, 76–78, 79, 80, 86, 89–90, 102, 117, 145

Sagan, Carl
  Contact, 133

Saturn, 32, 38

Schramm, David, 175

Schrödinger, Erwin, 91, 92, 95, 97, 98, 102, 181

Schwarzschild, Karl, 144–45, 149, 152–53

Schwarzschild black hole, 153

Schwarzschild radius, 149, 151, 157

Schwinger, Julian, 97, 108–10

Science, 8, 158–59

Science fiction, ix, 45, 131, 132, 133, 135

Scientific revolution, x, 42–43

Seneca, 25

Shapely, Harlow, 162–64

Singularity(ies), 153, 157, 206
  big bang, 180–82
  existence of, 154–55

Snyder, Hartland, 145

Socrates, 9–10, 22

Socratic method, 10n

Sodium atoms, 73

Solar system, 38, 50

Earth at center of, 43, 47
  sun at center of, 35–36, 44

Solar wind, 150

Solids, 55, 59, 191

Somnium (Kepler), 45

Sound
  and motion, 164

Sound waves, 124, 164

Space, 9, 26, 171, 180, 185, 187–88
  concept of, 207
  creation of, 29
  curvature of, 159, 169, *178*, 179
  cyclical, 55
  dimensionality of, 131, 192
  dimensions of, xi, 114–15, 181, 182, 187, 190
  distortions of, 148
  end of, 182, 206
  as fundamental quantity, 155
  in general relativity, 133
  infinite, 38
  infinitely compressed at beginning of big bang, xi

Space travel, 131, 156–57

Space warp, 131, 132

Spaces
  in mathematics, 6n

Spacetime, 148, 155, 182, 207

Spacetime bubbles, 184, 187, 189

Space-time dimensions, 114–15

"Spaghettification," 151

Special theory of relativity, 95–96, 105, 108, 120–21, 122, 125, 130, 137, 148
  in quantum field theory, 97–99
  infinity barrier in, 139

Spectrum, 89n

Speculation, 184, 185
  metaphysical, 208

Speed of light, 105, 120, 150
  energy required to accelerate to, 128

Speed of light, (*cont'd*)
  and infinite mass, 151
  as a limiting velocity, 129, 130
  with respect to motion of earth, 123–24, 148
  same for all observers, 124–25
  travel at, 122
Spherical mass, 144
Spiral nebulae, 160, 161, 163–64
Star Trek, 131
Star Trek: Deep Space Nine, 134
Stars, 38, 48–49, 191, 192
  brightness of, 162
  classified according to color, 162
  clustered together in galaxies, 51–52
  death of, 143, 148
  distance from earth, 37–38
  mass contained in, 168
  measuring distances of, 159–62
  nuclear reactions in, 140
Stellar collapse, 142, 145, 146, 147, 149, 151, 152, 153, 155
  gravitational fields surrounding, 150
  process of, 152
Stellar parallax, 37, 43, 160
Stern, Otto, 89
Stoics/Stoicism, xi, 23–25, 30, 32, 55, 204–05
  concept of time, 25–26, 28, 40
Stoney, George Johnstone, 74
Strong forces, 111–13, 191
Subatomic particles, 19, 92, 105, 128, 170
  behavior of, 91, 95
  energy of, 173
  properties of, 201–02
Subatomic phenomena, 102
Sudarshan, George, 137, 138
Summa Theologica (Aquinas), 18–19
Sun (the), 32, 33, 34, 40, 43, 121

center of solar system, 35–36, 43–44
death of, 140
gravitational fields surrounding, 150
motion of, 34, 37
rotation of, 153
surface temperature of, 162
Superconducting Supercollider, 188*n*
Supernova explosions, 142–43, 149, 192
Superstring theory, 113, 115–16, 119, 155, 201, 206
  probability of being true, 183
Synchrotron radiation, 87

Tachyons, 137–39
Telescope, 37, 41–42, 163
  of Galileo, 46–48
  Mount Wilson, 163
Tempier, Etienne, 28
Theoretical physics/physicists, 97, 98, 135, 137, 138, 200
Theory(ies)
  explaining forces of nature, 113–15, 116–17
  infinite quantities indicate something wrong with, 71, 150–51
  and speculation, 182–83
Theory creation, 81, 98
Theory of relativity, ix, 123, 125
  discovery of, 129–30
  forbids faster-than-light travel, 128
Thermodynamics, 83
Thompson, Sylvanus
  Calculus Made Easy, 68–69
Thomson, J. J., 74–75, 76, 77, 79, 85–86, 145
  Conduction of Electricity through Gases, 86
Thomson, William
  see Kelvin, Lord
't Hooft, Gerard, 136

Thorne, Kip, 133–34, 187

Tidal forces/tides, 48, 154, 157

Timaeus (Plato), 29

Time, 9, 29, 180, 187–88, 207
  changing perspectives in, 124–26, 133, 135
  circular, 20–21, 27
  cyclical, 205
  dimension of, xi, 114, 181, 190
  end of, 182, 206
  energy and, 103–04
  and eternity, 28–29
  finiteness of, 19–20
  as fundamental quantity, 155
  in general relativity, 133
  infinite, 17–31, 40
  nature of, 29
  passage of, 14–15
  Stoic conception of, 25–26, 28, 40
  see also Imaginary time; Linear time

Time machines, 133, 135–37

Time perspective
  distortion of, 148, 149, 151, 152, 155

Time travel, 120, 127–28, 130, 133, 135–37
  paradoxes associated with, 137–38
  possibility of, 139

Tinsley, Beatrice, 175

Tomonaga, Shin'ichiro, 97, 110

Totalitarian Rule of Physics, 138

Transfinite numbers, 5–7, 14, 70

Trinity, 61

Trojan War, 22

Tryon, Edward, 186, 187

Ultraviolet catastrophe, 80–82, *82*, 83, 88, 150–51

Ultraviolet light, 164–65

Ultraviolet radiation, 83

Unbearable Lightness of Being, The (Kundera), 30

Uncertainty principle, 103–04, 129

Unified theory, 113

Universe
  age of, 71, 72, 169–71, 177
  beginning of/creation of, 53, 134, 169–71, 180–82, 186–88
  as big nothing, 185–86
  borderline between open and closed, 159, 176, 179
  character of, 174–75, 176
  circumference of, 115
  clumpy structure of, 51–52
  dimensions of, 200
  Einsteinian conception of, 159
  expanding/contracting, 53, 72, 164–67, *168*, 173–74, 191
  fate of, 173–75
  finite/infinite, 31, 32–33, 38–39, 40, 48–49, 51–53, 71–72, 158–83, 187, 205
  God and, 196–97
  has no boundaries, 165, 171, 187
  "histories" of, 198–99
  hospitable to life, 190–92
  initial conditions of, 197, 198
  nature of, x
  rate of expansion of, 169–70
  static, 166, 167, 174
  structure of, 55
  theories explaining origin of, 95, 180–82, 186–88

Universes
  hypothetical, 191–92
  infinite number of (proposed), ix, 184–85, 197, 198, 199, 207–08
  open/closed, 167–69, *168*, 173–74, 175–77, 180, 182
  other, 184, 185, 187–88, 189–90, 195–96, 200, 202
  self-reproducing, 189
  see also Alternate universes

Unmoved Mover, 19, 20
Unreason, xi, 13–14

Vacuum, 200
Variable stars, 161
Velocity(ies)
    average, 58
    instantaneous, 58–59, 62, 63, 64,
        67–68
    of falling objects, 56, 57, 58–59
Velocity of light
    see Speed of light
Venus, 32, 46
Vibrating-electron model, 82
Vilenkin, Alexander, 187, 189
Virtual electrons, 104
Virtual particles, 103–04, 106–07, 110,
    129, 134, 185, 187, 200, 205
Virtual photons, 104
Virtual positrons, 104
Virtue, 24, 26
Viviani, Vincenzo, 56
Void, infinite, 26, 32–33, 40, 203,
    205
Volkoff, George, 145
Voltaire, 66
Von Laue, Max, 89
Vortex-ring theory/vortex rings, 73, 76

W particle, 112, 113
Wallis, John, 60, 64
    Arithmetic Infintorum, 203
Water, 24
    properties of, 194–96
Wave mechanics, 91
Wavelengths, 81, 150, 151, 164, 165,
    194
    calculation of, 94
    energy emission at, 82–83
    and temperature, 81

Waves, 84, 102
Waves of probability, 92
Weak forces, 111–13, 191
Weierstrass, Karl, 68
Westfall, Richard, 50
Weyl, Hermann, 114
"What the Tortoise Said to Achilles"
    (Dodgson), 17–18
Wheeler, John, 145–46, 206–07
White dwarfs, 140, 141, 142, 145
Whitrow, G. J.
    Natural Philosophy of Time, The, 13
"Why There Is Nothing Rather than
    Something" (Coleman), 200
Wien, Wilhelm, 83, 90
Wien's law, 83
Wilson, Robert, 170
World
    creation of, 26, 28, 40
    hallucinatory nature of, xv
Worlds
    infinite, 32–53
    other, 38–42
Wormholes, 133–35, 154, 185, 187
    macroscopic, 135
    microscopic, 200, 201–02
Wren, Christopher, 50

X rays, 73, 76, 83, 87, 89, 147

Yin and yang, 21
Yurtsever, Ulvi, 133–34

Z particle, 112, 113
Zeno of Citium, 23–24
Zeno of Elea, 8–11, 23, 203–04
    death of, 14–15
    paradoxes of, x, 7–8, 9, 11–12,
        13–14, 17, 59, 67, 204
Zwicky, Fritz, 175